普通高等教育"十二五"规划教材

C++程序设计实践指导

主　编　梁凤兰　郑步芹
副主编　史洪玮　于启红

东南大学出版社
·南京·

内 容 简 介

本教程是以 Visual C++ 6.0 为开发平台。全书共分为两个部分:第一部分为C++面向对象的基础实验部分;第二部分为课程设计部分。其中,第一部分给出了 10 个基础实验,基本覆盖了C++面向对象程序设计的主要知识点、方法和技巧;第二部分给出了可供课程设计讲解的学生成绩管理系统、通讯录管理系统和学生选课系统三个实际应用项目案例。

本书可以作为各类高等院校、高职院校C++面向对象程序设计课程的实践指导书,也可以作为读者自学用书。

图书在版编目(CIP)数据

C++程序设计实践指导 / 梁凤兰,郑步芹主编. —南京:东南大学出版社,2015.12
 ISBN 978-7-5641-5898-9

Ⅰ. ①C… Ⅱ. ①梁… ②郑… Ⅲ. ①C语言—程序设计—高等职业教育—教学参考资料 Ⅳ. ①TP312

中国版本图书馆 CIP 数据核字(2015)第 283957 号

C++程序设计实践指导

出版发行:	东南大学出版社
社　　址:	南京市四牌楼 2 号　邮编:210096
出 版 人:	江建中
网　　址:	http://www.seupress.com
经　　销:	全国各地新华书店
印　　刷:	南京玉河印刷厂
开　　本:	787mm×1092mm　1/16
印　　张:	10.25
字　　数:	237 千字
版　　次:	2015 年 12 月第 1 版
印　　次:	2015 年 12 月第 1 次印刷
印　　数:	1—3000 册
书　　号:	ISBN 978-7-5641-5898-9
定　　价:	22.00 元

本社图书若有印装质量问题,请直接与营销中心联系。电话:025-83791830

　　C++面向对象实现了类的封装、数据隐藏、继承、多态,使得其代码易维护及可重用,是高校计算机专业重要的学习内容之一,同时C++面向对象程序设计是一门实践性很强的课程,上机实验和课程设计是其不可缺少的实践环节。实践的目的是帮助学生加深和巩固对理论知识的理解,培养学生学习的兴趣和编写程序的信心,真正能运用C++这个强大、高效的编程工具去解决实际问题。

　　本书共分为两个部分:第一部分为C++基础实验,共有10个实验,每个实验都含有实验目的、实验内容和实验分析三个内容,基本覆盖C++程序设计的每个知识点,是学习C++程序设计首先要掌握的内容,这些内容对于加强学生基本功的训练,让学生打下扎实的基础是必须的。第二部分为C++课程设计,在此安排了学生成绩管理系统、通讯录管理系统和学生选课系统三个实训项目。学生成绩管理系统给出了详细的项目设计流程和源程序代码,重点介绍了利用面向对象的知识封装链表操作完成管理系统中增删改查的方法。通讯录管理系统和学生选课系统给出功能设计、完整的源程序代码和详细的注释。通过这些项目训练可以帮助学生系统掌握C++主要内容,深刻理解课本理论知识,进一步掌握面向对象的程序设计方法,培养学生实际分析问题和动手的实践能力。

　　本书所有参编人员都是长期在高校从事专业教学与科研的一线教师,具有丰富的编程与教学经验,了解在学习过程中要求和希望掌握的知识以及容易出错的地方,并在书中相应的地方予以较详细的讲解。本书第一部分C++基础实验由梁凤兰编写,本书第二部分C++课程设计由郑步芹编写,于启红和史洪玮完成了本书的校对工作。

　　由于编者水平和经验所限,书中不足和错误之处在所难免,恳请广大读者批评指正。

编者
2015.6

第一部分　C++基础实验 1

实验一　C++对C的扩充 2
实验二　类和对象(一) 10
实验三　类和对象(二) 17
实验四　类和对象(三) 22
实验五　运算符重载 30
实验六　继承和派生(一) 40
实验七　继承和派生(二) 49
实验八　多态性与虚函数 58
实验九　输入输出流 72
实验十　异常处理 84

第二部分　C++课程设计 92

样例一　学生成绩管理系统 93
样例二　通讯录管理系统 115
样例三　学生选课系统 129

第一部分　C++基础实验

　　学习程序设计语言的主要目的是能够利用简洁的语句编写出高效、完整的实用程序,以解决在各个方面遇到的具体问题。在掌握了C++语言的基本概念及语法后,还需要进一步掌握面向对象程序方法及各种编程技巧,才能真正成为一个软件编程人员。

　　本课程是软件工程、计算机专业的一门专业课,教材采用 Microsoft Visual C++ 6.0 作为程序运行的环境。Visual C++由于其界面友好、操作方便等优点,是从事计算机程序设计的最佳编程工具之一,同时也是学习 Windows 编程的极好入门语言。通过本实验的学习,使学生掌握用面向对象程序设计语言中常用到的概念和术语以及面向对象程序设计的特点:封装性、继承和派生性以及多态性等,为其他编程工具的学习以及成为一名熟练的程序员打下坚实的基础。

　　实验一,通过本次实验主要掌握的内容是C++对 C 的扩充,主要有输入/输出、函数的重载、函数原型声明、有默认参数的函数等。

　　实验二、实验三、实验四,通过这三个实验主要掌握声明和对象的含义、掌握构造函数和析构函数的特点、掌握构造函数和析构函数的调用顺序和了解友元函数。

　　实验五,通过本次实验主要掌握运算符重载的方法、掌握运算符重载的特点、掌握单目和双目运算符重载、掌握自定义类型和标准类型间的相互转换。

　　实验六、实验七,通过这两个实验主要掌握派生类的含义和声明以用派生类各成员访问属性、掌握派生类构造函数和析构函数以及它们的调用顺序和掌握基类与派生类的转换。

　　实验八,通过本次实验主要掌握多态性的含义、掌握虚函数的定义和作用场合、了解纯虚函数和抽象类的概念。

　　实验九,通过本次实验主要理解标准输入输出流类的作用,掌握标准输入输出流对象的使用、掌握文件打开关闭的函数。

　　实验十,通过本次实验主要掌握异常处理的任务、掌握异常处理的方法。

实验一 C++ 对 C 的扩充

一、实验目的

1. 进一步熟悉在 Microsoft Visual C++ 6.0 的上机环境中编译、连接和运行 C++ 程序的方法。

2. 掌握 C++ 对 C 扩充了哪些功能,并善于在编写程序过程中应用这些新功能。具体要求掌握以下知识:

掌握 C++ 中的输入/输出流对象的作用;

掌握函数原型的形式和使用场合;

掌握函数重载的概念、条件并能应用到程序中;

掌握默认参数函数的参数如何赋默认值以及和函数重载的区别;

掌握变量引用的概念的相关知识和引用作为函数参数的使用方法,以及值传递、地址传递和引用传递的区别;

理解 const 定义常变量和宏定义的区别;

理解内置函数的含义和应用的场合;

掌握作用运算符的简单使用;

掌握 new 和 delete 运算符的使用,理解动态分配/撤销内存的含义,理解动态和静态的区别。

3. 进一步熟悉 C++ 程序的结构和编程方法。

二、实验内容

1. 输入以下程序,进行编译、观察编译情况,如果有错误,请修改程序,再进行编译,直到没有错误,然后进行连接和运行,分析运行结果。

```
#include <iostream>
using namespace std;
int main()
{
    int a,b;
    c=add(a,b);
    cout<<"a+b="<<c<<endl;
    return 0;
}
```

```
int add(int x,int y);
{
    int z;
    z=x+y;
    return(z);
}
```

2. 有以下程序,请完成下面工作。

(1) 阅读程序,写出运行时输出的结果;

(2) 上机运行,验证结果是否正确;

(3) 用单步调试分析程序执行过程,尤其是各函数的调用过程中,实参和形参传递的方式。

```
#include <iostream>
using namespace std;
void swap1(int a,int b)
{
    int t;t=a;a=b;b=t;
}
void swap2(int *p1,int *p2)
{
    int t;t=*p1;*p1=*p2;*p2=t;
}
void swap3(int *p1,int *p2)
{
    int *t;t=p1;p1=p2;p2=t;
}
void swap4(int &r1,int &r2)
{
    int t;t=r1;r1=r2;r2=t;
}
int main()
{
    int m=10,n=20;
    swap1(m,n);      cout<<"m="<<m<<",n="<<n<<endl;
    swap2(&m,&n);    cout<<"m="<<m<<",n="<<n<<endl;
    swap3(&m,&n);    cout<<"m="<<m<<",n="<<n<<endl;
    swap4(m,n);      cout<<"m="<<m<<",n="<<n<<endl;
    return 0;
```

}

3. 编写函数求圆和长方形的面积。要求用函数重载实现。

4. 编写一个程序，用来求2个或3个正整数中的最大数。要求用带有默认参数的函数实现。

三、实验分析

1. 将程序输入到 V C++ 6.0 的运行环境中，编译时显示如图 1.1.1 所示的错误。第6行有2个错误，第1个错误是变量c没有定义，只需要增加变量c的定义即可，也就是将"int a,b;"改为"int a,b,c;"；第2个错误是关于"函数原型"的问题，在C++中规定，如果函数调用的位置在函数定义之前，则要求在函数调用之前必须对所调用的函数作函数原型说明，函数原型说明的形式"函数类型　函数名(函数参数列表);"，所以在函数调用前加上"int add(int x,int y);"，再次编译时，第10行的错误随之消失；第11行的错误是"int add(int x,int y);"语句后多了";"，函数定义时函数首部的后面是不需要加";"的。再次编译时将显示如图1.1.2所示的编译结果，这2个警告，表示变量a,b在使用前没有初始化，所以应该在主函数中增加输入语句"cout<<"请输入两个整数 a b 的值:";cin>>a>>b;"，再次编译程序，此时可以运行程序，程序的运行结果如图1.1.3所示。

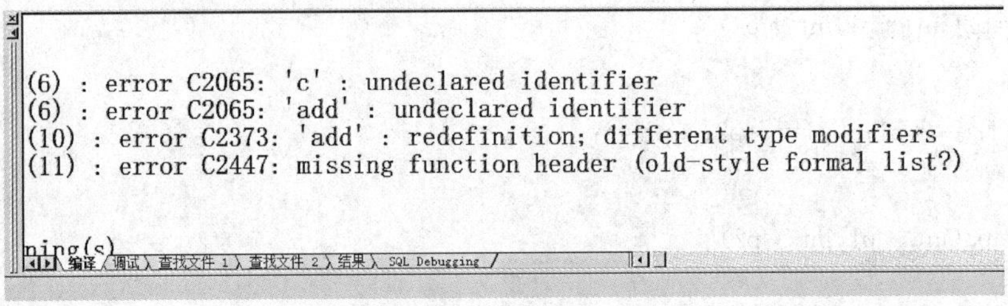

图 1.1.1　编译结果图

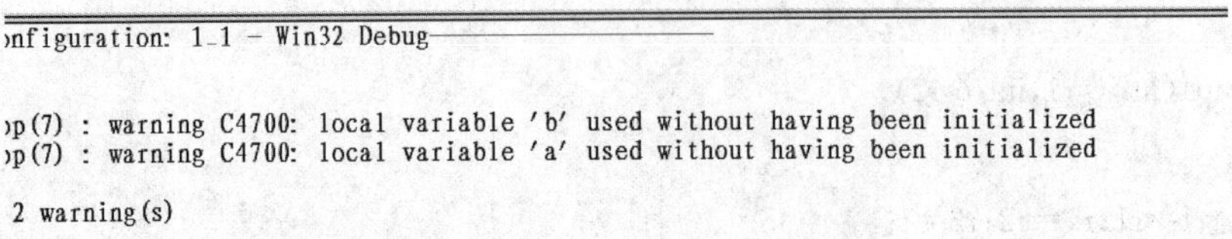

图 1.1.2　编译结果图

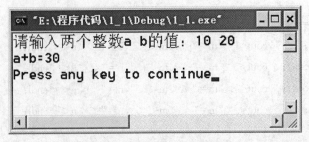

图 1.1.3　程序运行结果图

2. 本题主要考查函数参数传递的形式即"值传递、地址传递和引用传递",其中函数 swap1 为值传递,函数 swap2 和 swap3 为地址传递,函数 swap4 为引用传递,各函数具体调用过程如下:

(1) 函数 swap1 的调用过程

①参数传递:int a=m,int b=n,将实参的值传递给形参,即将 m 的值传递给 a,n 的值传递给 b 如图 1.1.4 所示。

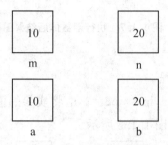

图 1.1.4　参数传递结果图

②执行函数体:t=10;a=20;b=10;即 a,b 的值互换,如图 1.1.5 所示。

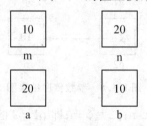

图 1.1.5　执行函数体后结果图

③返回到调用函数的地方。

(2) 函数 swap2 的调用过程

①参数传递:int * p1=&m,int * p2=&n,将实参的值传递给形参,即将 m 的地址传递给 p1,n 的地址传递给 p2,如图 1.1.6 所示。

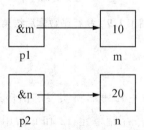

图 1.1.6　参数传递结果图

②执行函数体:t=10;*p1=20;*p2=10;即 m,n 的值互换,如图 1.1.7 所示。

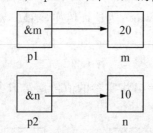

图 1.1.7　执行函数体后结果图

③返回到调用函数的地方。

(3) 函数 swap3 的调用过程

①参数传递:int * p1=&m,int * p2=&n,将实参的值传递给形参,即将 m 的地址传递给 p1,n 的地址传递给 p2,如图 1.1.8 所示。

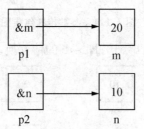

图 1.1.8　参数传递结果图

②执行函数体:t=&m;p1=&n;p2=&m;即 p1,p2 的值互换,如图 1.1.9 所示。

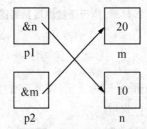

图 1.1.9　执行函数体后结果图

③返回到调用函数的地方。

(4) 函数 swap4 的调用过程

①参数传递:int &r1=m,int &r2=n,将实参的值传递给形参,即 r1 是 m 的引用,r2 是 n 的别名,也就是 r1 和 m 占用同一存储单元,r2 和 n 占用同一存储单元,如图 1.1.10 所示。

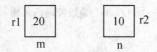

图 1.1.10　参数传递结果图

②执行函数体:t=20;r1=10;r2=20;如图1.1.11所示。

图 1.1.11　执行函数体后结果图

③返回到调用函数的地方。

程序运行结果如图 1.1.12 所示。

图 1.1.12　程序运行结果图

3. 本题考查的是函数重载,只有当函数名相同,而函数的参数个数不同或参数的类型不同时才能实现函数的重载;但只有函数的返回值类型不同时是不能实现函数重载的。程序代码如下:

```
#include <iostream.h>
const float PI=3.14;
//计算长方形的面积,有2个参数
float area(float a,float b)
{
    return a*b;
}
//计算圆的面积,有1个参数
float area(float r)
{
    return PI*r*r;
}
int main()
{
    float a,b,r;
    cout<<"请输入圆的半径:";
    cin>>r;
    cout<<"请输入长方形的长、宽:";
```

```
cin>>a>>b;
cout<<"圆的面积是:"<<area(r)<<endl;
cout<<"长方形的面积是:"<<area(a,b)<<endl;
return 0;
}
```

程序运行结果如图1.1.13所示。

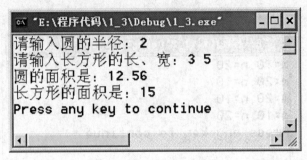

图1.1.13　程序运行结果图

计算长方形面积area函数有2个参数,计算圆面积area函数有1个参数,满足函数重载的函数名相同,但参数个数不同的条件,所以可以实现函数重载。

4. 本题考查的是有默认参数的函数,题目要求的是"正整数中的最大数",因为任何一个正整数都比0大,所以可以把最后一个形参设默认值为0。程序代码如下:

```
#include <iostream.h>
int max( int a, int b, int c=0)
{
    int d;
    d=a>b? a:b;
    return d>c? d:c;
}
int main()
{
    int a,b,c;
    cout<<"请输入三个正整数:";
    cin>>a>>b>>c;
    while(a<=0||b<=0||c<=0)
    {
        cout<<"输入有误,请重新输入:";
        cin>>a>>b>>c;
    }
    cout<<"两个数中较大值是:"<<max(a,b)<<endl;
```

cout<<"三个数中最大值是:"<<max(a,b,c)<<endl;
return 0;
}

程序运行结果如图 1.1.14 所示。

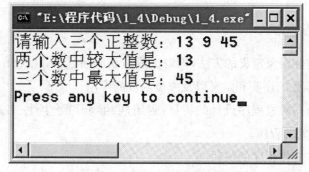

图 1.1.14　程序运行结果图

实验二 类和对象(一)

一、实验目的

1. 掌握声明类和对象定义的方法,理解类成员访问属性 public、protected 和 private 的含义,掌握类成员函数的声明和定义以及成员函数的作用。
2. 初步理解面向对象程序设计的特征:封闭性、继承和派生性、多态性,初步掌握用类和对象编制基于对象的程序。
3. 初步掌握多文件程序的编写。
4. 初步掌握调试基于对象的程序。

二、实验内容

1. 有以下程序:

```cpp
#include <iostream>
using namespace std;
class Time
{
public:
    int hour;
    int minute;
    int sec;
};
int main()
{
    Time t1;
    Time &t2=t1;
    cin>>t2.hour;
    cin>>t2.minute;
    cin>>t1.sec;
    cout<<t1.hour<<":"<<t1.minute<<":"<<t2.sec<<endl;
    return 0;
}
```

改写程序,要求:
(1) 将数据成员改为私有的;

(2) 将输入和输出的功能改为由成员函数实现;

(3) 在类体内定义成员函数。

2. 求3个长方体体积,请编一个基于对象的程序。数据成员包含 length(长)、width(宽)、height(高)。要求用成员函数实现以下功能:

(1) 由键盘输入3个长方体的长、宽、高;

(2) 计算长方体的体积;

(3) 输出3个长方体的体积。

要求:(1) 将类的定义放在头文件 box.h 中;

(2) 将成员函数定义放在源文件 box.cpp 中;

(3) 主函数放在源文件 main.cpp 中。

3. 建立一个数组类 Arr,根据已知数组 a 的元素值产生新数组 b。产生规则是:数组 b 的任一元素的值是数组 a 对应元素及其后连续两个元素的平均值,即 b[i]=(a[i]+a[i+1]+a[i+2])/3。假定数组最后一个元素的后续元素为第0个元素,即若数组有 n 个元素,最后一个元素是 a[n−1],而 a[n−1]的后续元素是 a[0]。最后输出数组 b 的各个元素值及元素值的获取规则。具体要求如下:

(1) 私有数据成员

● int a[100]:初始数组。

● double b[100]:生成的新数组。

● int n:数组元素个数。

(2) 公有成员函数

● void init(int t[],int n1):初始化函数,用 t 初始化数组 a,用 n1 初始化 n。

● void fun():按规则生成数组 b。

● void print():输出数组 b。要求每行输出一个元素值,同时输出该元素的产生规则,即该值是哪三个数值的平均值。

(3) 在主函数中定义一个具有10个元素的整型数组 data,其初值是{2,4,6,8,10,12,14,16,18,20}。定义一个 Arr 类的对象 ar,用 data 数组及其元素个数初始化 ar。通过 ar 调用成员函数,产生并输出新数组 b 的各个元素值及元素值的获取规则。本题正确的输出结果为

b[0]=4=(2+4+6)/3.0
  b[1]=6=(4+6+8)/3.0
    …
      b[8]=13.3333=(18+20+2)/3.0
      b[9]=8.66667=(20+2+4)/3.0

三、实验分析

1. 该题主要考查类声明和对象引用的相关知识,将数据改为私有成员后,类的声明

如下：
```
class Time
  {private:
  int hour;
  int minute;
  int sec;
  };
```
将输入和输出的功能改为由成员函数实现，并在类体内定义成员函数，类的声明如下：
```
class Time
{
private:
    int hour;
    int minute;
    int sec;
public:
    void input()
    {
      cout<<"请输入 hour minute sec";
      cin>>hour>>minute>>sec;
    }
    void output()
    {
      cout<<hour<<":"<<minute<<":"<<sec<<endl;
    }
};
```
主函数相应改为：
```
int main()
{
    Time t1;
    Time &t2=t1;
    t1.input();
    t1.output();
    t2.output();
    return 0;
}
```
一般情况下，将需要被外界调用的成员函数指定 public 访问权限，它们是类的对外接口，但要注意，并非要把所有成员函数都指定为 public 访问权限，有的函数并不是准备为外界调用的,而是为本类中的成员函数所调用的,就应该将它们指定为 private 访问权限;成员函数体代码比较少(2~3行),一般可在声明类时在类体中定义,函数体代码比较多(多于3行),一般在类体内声明函数原型,在类外定义。将数据成员指定为 private 的访问权限。程序运行结果如图 1.2.1 所示。

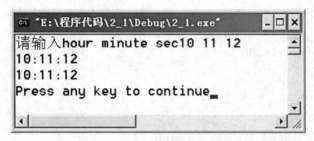

图 1.2.1 程序运行结果图

2. 该题考查的是多文件的程序。如果一个类只被一个程序使用,那么类的声明和成员函数的定义直接写在程序的开头,但是如果一个类被多个程序使用,这样做的重复工作量就很大,效率就太低了。在面向对象的程序开发中,一般做法是将类的声明(包含成员函数的声明)放在指定的头文件中,用户如果想用该类,只要把头文件包含进来即可,不必在程序中重复书写类的声明。

以下代码是头文件 box.h 中的代码,是 Box 类的声明,包含成员函数的声明。

```
class Box
{
private:
    int length,width,heigth,v;
public:
    void init(int l,int w,int h);
    void volume();
    void print();
};
```

以下代码是源文件 box.cpp 中的代码,是 Box 类中成员函数的定义,"#include <iostream.h>"和"#include "box.h""两个文件包含,前者用的是尖括号,尖括号表示只在系统默认目录的路径查找,通常用于包含系统中自带的头文件,后者用的是双撇号,编译器从用户的工作目录开始搜索。如果未找到则到系统默认目录查找,通常用于包含程序用户编写的头文件。

```
#include <iostream.h>
#include "box.h"
void Box::init(int l,int w,int h)
{
```

```
        length=l;
        width=w;
        heigth=h;
}
void Box::volume()
{
        v=length*width*heigth;
}
void Box::print()
{
        cout<<"长方体的体积是:"<<v<<endl;
}
```

以下代码是源文件 main.cpp 中的代码,是 main 函数的定义。

```
#include <iostream.h>
#include "box.h"
int main()
{
    int h,l,w;
    cout<<"请输入长方体的长、宽、高:"<<endl;
    cin>>l>>w>>h;
    Box b1;
    b1.init(l,w,h);
    b1.volume();
    b1.print();
    return 0;
}
```

程序运行结果如图 1.2.2 所示。

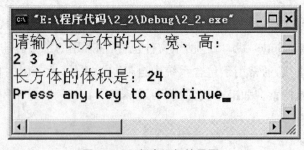

图 1.2.2　程序运行结果图

3. 根据题意的描述,类 Arr 声明的代码如下:

　　#include <iostream.h>

```cpp
class Arr
{
private:
    int n,a[100];
    double b[100];
public:
    void init(int t[],int n1);
    void fun();
    void print();
};
```

类 Arr 中成员函数定义的代码如下：

```cpp
void Arr::init(int t[],int n1)
{
    n=n1;
    for(int i=0;i<n;i++)
    {
        a[i]=t[i];
    }
}
void Arr::fun()
{
    for(int i=0;i<n;i++)
        b[i]=(a[i]+a[(i+1)%10]+a[(i+2)%10])/3.0;
}
void Arr::print()
{
    for(int i=0;i<n;i++)
    {
        cout<<"b["<<i<<"]="<<b[i];
        cout<<"=("<<a[i]<<"+"<<a[(i+1)%10]<<"+"<<a[(i+2)%10];
        cout<<")/3.0"<<endl;
    }
}
```

主函数的定义如下：

```cpp
int main()
{
    int data[10]={2,4,6,8,10,12,14,16,18,20};
```

Arr arr1；

arr1.init(data,10)；

arr1.fun()；

arr1.print()；

return 0；

}

程序运行结果如图1.2.3所示。

图1.2.3　程序运行结果图

实验三　类和对象(二)

一、实验目的

1. 理解构造函数和析构函数的作用,掌握构造函数和析构函数的特点以及两者的区别,掌握构造函数的重载和使用默认参数的构造函数。
2. 掌握构造函数和析构函数的调用顺序。
3. 理解对象数组、对象指针的含义和作用,特别掌握 this 指针的含义和作用。

二、实验内容

1. 有以下程序,请完成下面工作
(1) 阅读程序,写出运行时输出的结果;
(2) 上机运行,验证结果是否正确;
(3) 用单步调试分析程序执行过程,尤其是构造函数和析构函数的调用顺序。

```
#include <iostream>
using namespace std;
class  Sample
{
private:int   a,b;
public:
    Sample()
    {
      a=0;
      b=0;
      cout<<"调用了构造函数   a="<<a<<", b="<<b<<endl;
    }
    Sample(int x)
    {
      a=x;
      b=0;
      cout<<"调用了构造函数   a="<<a<<", b="<<b<<endl;
    }
    Sample(int x,int y)
    {
```

```
            a=x;
            b=y;
            cout<<"调用了构造函数  a="<<a<<", b="<<b<<endl;
        }
        ~Sample()
        {
            cout<<"调用了析构函数  a="<<a<<", b="<<b<<endl;
        }
};
int  main( )
{
    Sample s1,s2(10),s3(20,30);
    return 0;
}
```

2. 有以下程序,请完成下面工作

(1) 阅读程序,写出运行时输出的结果;

(2) 上机运行,验证结果是否正确;

(3) 用单步调试分析程序执行过程,尤其是构造函数和拷贝构造函数的调用。

```
#include <iostream.h>
class Sample
{
private:int x;
public:
        Sample(int i = 10)
        {
            x=i;
            cout<<"Constructor called!   x="<<x<<endl;
        }
        Sample(Sample &a)
        {
            x=a.x;
            cout<<"Copy constructor called!  x="<<x<<endl;
        }
};
void main( )
{
```

```
        Sample    s1;
        Sample    s2 = s1;
}
```

3. 设计一个类,用来表示直角坐标系中的任意一条直线并输出它的属性。

三、实验分析

1. 程序中定义了一个 Sample 类,该类有三个构造函数,这三个构造函数的参数不同所以构成函数的重载。在执行"Sample s1,s2(10),s3(20,30);"语句时,首先创建对象 s1,此时去调用无参的构造函数,所以输出结果为如图 1.3.1 所示的第一行;其次创建对象 s2,此时去调用有一个参数的构造函数,所以输出结果为如图 1.3.1 所示的第二行;再次创建对象 s3,此时去调用两个参数的构造函数,所以输出结果为如图 1.3.1 所示的第三行;程序再向下执行时,s1,s2,s3 的生存周期就要结束了,在收回对象占用存储空间前要对所占用的存储空间做一些清理工作,该清理工作由析构函数完成,所以要调用 Sample 类的析构函数,根据"先构造的后析构,后构造的先析构"原则,所以首先调用对象 s3 析构函数,输出结果如图 1.3.1 所示的第四行;其次调用对象 s2 析构函数,输出结果如图 1.3.1 所示的第五行;最后调用对象 s1 析构函数,输出结果如图 1.3.1 所示的第六行。

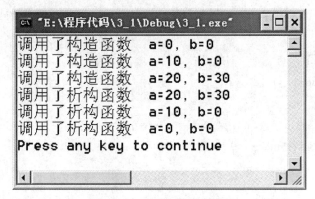

图 1.3.1 程序运行结果图

2. 程序中类 Smaple 中定义了两个构造函数,一个是默认参数构造函数,另一个是拷贝构造函数。程序在执行"Sample   s1;"语句时调用默认参数构造函数,因为调用构造函数时没给出实参的值,所以参数的值即为默认形参的值,即数据成员 x 的取值为 10,所以输出的结果如图 1.3.2 所示的第一行;程序在执行"Sample   s2=s1;"时,该语句可以等价写为"Sample s2(s1);"该语句中调用构造函数的参数是一个对象,所以应该调用拷贝构造函数,执行"x=a. x"语句时,相当于"x=s1. x",因为 a 是 s1 的别名,所以对象 s2 数据成员 x 的值为 10,所以输出的结果如图 1.3.2 所示的第二行。

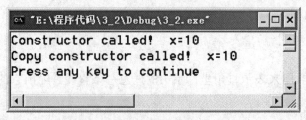

图 1.3.2 程序运行结果图

3. 直线类的属性有直线的两个点和直线的长度,类的描述如下:

直线类

{数据成员;//用于表示直线的两个点

构造函数;//用于初始化两个点

成员函数 1;//用于输出直线的两个点

成员函数 2;//用于输出直线的长度

};

具体实现的代码如下:

```
#include <iostream.h>
#include <math.h>
class Line
{
        private:
            int x1, y1, x2, y2;
        public:
            Line(int =0, int =0, int =0, int =0);
            void printPoint();
            double getLength();
};
inline Line::Line(int a, int b, int c, int d)
{
    x1 = a;
    y1 = b;
    x2 = c;
    y2 = d;
}
inline void Line::printPoint()
{
    cout<<"A:"<< x1 <<", "<< y1 << endl;
    cout<<"B:"<< x2 <<", "<< y2 << endl;
```

}
inline double Line::getLength()
{
  double length;
  length = sqrt((x2－x1)*(x2－x1)+(y2－y1)*(y2－y1));
  return length;
}
void main()
{
  Line line(10,80,－10,12);
  line.printPoint();
  cout<< line.getLength() << endl;
}

程序运行结果如图 1.3.3 所示。

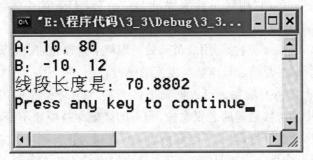

图 1.3.3 程序运行结果图

实验四 类和对象(三)

一、实验目的

1. 掌握对象的动态建立和释放,理解对象动态建立和静态建立的区别。
2. 掌握复制构造函数的作用和特点。
3. 掌握静态成员的特征和使用。
4. 理解友元的含义和友元带来的副作用,掌握将普通函数成员声明为类的友元函数的方法以及友元函数具有的特权。

二、实验内容

1. 定义一个 Dog 类,它用静态数据成员 Dogs 记录 Dog 的个体数目,静态成员函数 GetDogs 用来存取 Dogs。设计并测试这个类。

2. 声明复数 Complex 类,分别用成员函数实现两个复数的加法运算和用友元函数实现两个复数的减法运算,并总结成员函数与友元的区别。

3. 聪明的老鼠:有一只猫每天可以抓到很多老鼠,它在吃老鼠时习惯将抓到的所有老鼠排队,然后将其中的奇数位置的老鼠吃掉,剩下的老鼠保持顺序不变,仍然是吃掉其中的奇数位置上的老鼠,如此反复,直到只剩下最后一只老鼠为止。剩下的这只老鼠将会在第二天和新抓到的老鼠一起排队。然而,有一天这只猫发现一连几天幸存下来的都是同一只机灵的小老鼠。问这只聪明的小老鼠是如何避免被吃掉的命运的?

三、实验分析

1. 该题主要考查静态数据成员和静态成员函数。静态数据成员,在定义类时,如果希望某个数据为所有对象共享,就可以将该数据成员定义为静态数据成员,静态数据成员为所有对象共享,而不只属于某个对象的成员,所有对象都可以引用它。静态数据成员在内存中只占一份存储空间(而不是每个对象都分别为它保留一份空间),如果改变它的值,各对象中这个数据成员的值都同时改变了。静态数据成员只能在类体外进行初始化,不能用参数初始化表对静态数据成员初始化,如果未对静态数据成员初始化,则编译系统会自动赋值为 0。静态成员函数,在类中声明函数时,在函数类型前加 static 即为静态成员函数,静态成员函数与非静态成员函数的根本区别是:非静态成员函数有 this 指针,而静态成员函数没有 this 指针,因此静态成员函数不能访问本类中的非静态成员数据。具体程序代码如下:

```
#include <iostream>
using namespace std;
class Dog
```

```
{
private:
    static int dogs;//静态数据成员,记录Dog的个体数目
public:
    Dog(){}
    void setDogs(int a)
    {
        dogs = a;
    }
    static int getDogs()
    {
        return dogs;
    }
};
int Dog::dogs = 25;//初始化静态数据成员
void main()
{
    cout<<"未定义Dog类对象之前:x = "<< Dog::getDogs() << endl;;
                //x在产生对象之前即存在,输出25
    Dog a, b;
    cout<<"a中x:"<< a.getDogs() << endl;
    cout<<"b中x:"<< b.getDogs() << endl;
    a.setDogs(360);
    cout<<"给对象a中的x设置值后:"<< endl;
    cout<<"a中x:"<< a.getDogs() << endl;
    cout<<"b中x:"<< b.getDogs() << endl;
}
```

程序运行结果如图1.4.1所示。

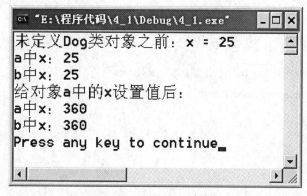

图1.4.1 程序运行结果图

2. 友元函数,在类中声明函数时,在函数类型前加 freind 即表明该函数为该类友元函数,友元函数虽然在类的内部声明,但并不是类的成员函数,因为函数和类成了朋友的关系,所以可以访问类的私有、受保护的成员。具体程序代码如下:

```cpp
#include <iostream.h>
class Complex
{
private:
    int real,imag;
public:
    Complex(int r=0,int i=0);
    Complex add(Complex c1);//成员函数实现复数的加法运算
    friend Complex sub(Complex c1,Complex c2);//用友元函数实现复数的减法运算
    void display();
};//声明复数类
Complex::Complex(int r,int i)
{
    real=r;
    imag=i;
}
Complex Complex::add(Complex c1)
{
    Complex c;
    c.real=real+c1.real;
    c.imag=imag+c1.imag;
    return c;
}
Complex sub(Complex c1,Complex c2)
{
    Complex c;
    c.real=c1.real-c2.real;//使用类中私有成员 real
    c.imag=c1.imag-c2.imag; //使用类中私有成员 imag
    return c;
}
void Complex::display()
{
    cout<<real;
    if(imag>=0)
```

```
        {
            cout<<"+"<<imag<<"i";
        }
        else
        {
            cout<<imag<<"i";
        }
        cout<<endl;
    }
    int main()
    {
        Complex c1(1,2),c2(3,4),c3,c4;
        c3=c2.add(c1);
        cout<<"c2+c1=";
        c3.display();
        c4=sub(c2,c1);
        cout<<"c2-c1=";
        c4.display();
        return 0;
    }
```

程序运行结果如图 1.4.2 所示。

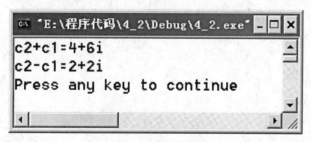

图 1.4.2　程序运行结果图

3. 第一种方法:假设每天新抓到的老鼠有 n 只,那么就会有 n+1 只老鼠进行排队。利用长度为 n+1 的整型数组分别代表 n+1 只老鼠,数组元素的值是 0 代表这只老鼠被吃掉了,是 1 代表这只老鼠在这一轮幸存下来了,被吃掉后的老鼠不占位置,只排队未被吃掉的老鼠,重复上述过程直到只剩下一只老鼠为止,该老鼠所在的数组幸免于难的位置就是所求的聪明的小老鼠避免被吃掉时应该占据的位置。

程序的实现流程可以描述如下,流程图如图 1.4.3 所示。

(1) 初始化数组,对 n+1 个老鼠进行排队;

(2) 对数组中值等于 1 的数组元素进行计数,将其中的计数值为奇数的数组元素的位

置 0,表示老鼠吃掉;

（3）重复上述过程,直到数组中值等于 1 的数组元素的个数只有一个为止;

（4）数组中值等于 1 的数组元素在数组中的位置,就是小老鼠在排队时避免被吃掉所占据的位置,输出该位置。

第二种方法:每次只是偶数位的老鼠会幸存下来,那么幸存下来的小老鼠的初始位置只能是 2 的 m 次幂的位置,并且是仅仅小于总的老鼠数 n+1 的 2 的 m 次幂,这样就可以求得 $m=\log_2(n+1)$,从而就可以求得聪明的小老鼠在排队时在什么位置会避免被吃掉。程序的实现流程可以描述如下:

（1）计算 m 的值,得到幸存老鼠的只数;

（2）输出该位置。

具体程序代码如下:

```
#include <iostream>
#include <math.h>
using namespace std;
class Cat_rat
{   public:
        int lookleft1(int);   //查找最后剩下的老鼠的初始位置
        int lookleft2(int);
private:
        int * rat;   //指向存放老鼠信息的数组的指针
        int count;   }   //指向老鼠报数的计数器
int Cat_rat::lookleft1(int m)
{
        int i,sum;
        rat=new int[m+1];
        for(i=0;i<=m;i++)   //给每个老鼠设置未报数的标记
}
            rat[i]=1;
}
```

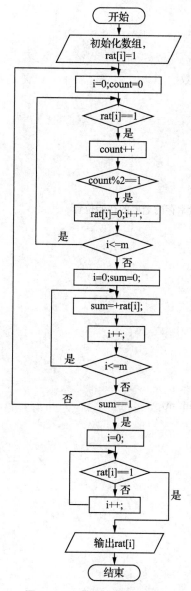

图 1.4.3 猫吃老鼠流程图

```
do   //剩余的老鼠报数
{
i=0;
count=0;
do
 {
      if(rat[i]==1){count++;}
      if(count%2==1){rat[i]=0;}
      i++;
```

```cpp
    }while(i<=m);
    i=0;
    sum=0;
    do   //查看剩余老鼠的个数
    {
         sum=sum+rat[i];
         i++;
       }while(i<=m);
    }while(sum!=1);
    for(i=0;;i++)    //查看幸存的老鼠的初始位置
    {
         if(rat[i]==1)
         {
             return i+1;
             break;
         }
        }
    }
    int Cat_rat::lookleft2(int m)
    {
    int n,i;
    n=m+1;
    i=0;
    do   //找到最大的 i 值
    {
       i++;
       n/=2;
       }while(n/2>0);
       n=1;
       while(i-->0)   //计算幸存老鼠的初始位置
       {n*=2;}
       return n;
    }
    int main()
    {
         int n;
         Cat_rat newday;
```

cout<<"猫每天捉到的新老鼠的个数为:";

cin>>n;

cout<<"总的老鼠个数为:"<<n+1<<endl;

cout<<"每天都能留下来的小老鼠所在的位置为:"<<endl;

cout<<"解法一:"<<newday.lookleft1(n)<<endl;

cout<<"解法二:"<<newday.lookleft2(n)<<endl;

return 0;

}

程序运行结果如图 1.4.4 所示。

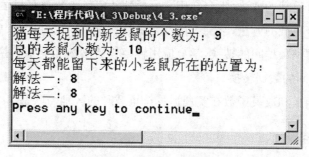

图 1.4.4　程序运行结果图

实验五　运算符重载

一、实验目的

1. 理解运算符重载的含义。
2. 掌握运算符重载的方法和运算符重载规则。
3. 通过运算符重载的方法,学会应用友元函数和成员函数对双目运算符的重载。
4. 通过运算符重载的方法,学会应用友元函数和成员函数对自增、自减运算符的重载。
5. 掌握不同类型数据间的转换,学会应用转换构造函数和类型转换函数实现标准类型和自定义类型数据间的数据转换。
6. 掌握友元函数和成员函数在实现运算符重载时的区别与联系。

二、实验内容

1. 定义一个复数类 Complex,重载"＋＋"运算符,使之能用于复数的自加即实部和虚部分别自加,分别用成员函数实现"＋＋"运算符前置运算和用友元函数实现"＋＋"运算符后置运算。

2. 定义一个复数类 Complex,重载"＋"和"－"运算符,使之能用于复数的加法和复数的减法运算,分别用成员函数实现复数对象加法运算和用友元函数实现复数对象减法运算。

3. 编写程序,重载运算符"＋"、"－"和"＊",对实数矩阵进行加法、减法和乘法运算,并且重载运算符"()",用来返回矩阵元素的值。

4. 设 d1、d2 为 double 类型数据,c1 为复数类型数据,现计算 d2＝d1＋c1,要求:计算"d1＋c1"时对"＋"运算符重载,将结果赋给 d2 时,用类型转换函数实现,即将复数类型转换成 double 类型,转换规则是将实部转换为 double 类型,虚部丢弃。

三、实验分析

1. 声明一个复数类 Complex,类中设置两个私有数据 real 和 imag,分别表示复数的实部和虚部,类中声明成员函数 Complex operator＋＋()用来实现"＋＋"运算符前置的重载,友元函数 friend Complex operator＋＋(Complex &c,int)用来实现"＋＋"运算符后置的重载。

类的声明如下:

```
#include <iostream.h>
class Complex
{
```

```cpp
    private:
        int real,imag;
    public:
        Complex(int r=0,int i=0);
        Complex operator++();
        friend Complex operator++(Complex &c,int);
        void display();
};
```

各函数定义如下：

```cpp
Complex::Complex(int r,int i)
{
    real=r;
    imag=i;
}
Complex  Complex::operator++()
{
    real++;
    imag++;
    return *this;
}
Complex operator++(Complex &c,int)
{
    Complex ct=c;
    c.real++;
    c.imag++;
    return ct;
}
void Complex::display()
    {
        cout<<real;
        if(imag>=0)
        {
            cout<<"+"<<imag<<"i";
        }
        else
```

```
        {
            cout<<imag<<"i";
        }
        cout<<endl;
    }
```

主函数定义如下：
```
int main()
{
    Complex c1(1,2),c2(3,4),c3,c4;
    c3=++c1;
    cout<<"c1=";c1.display();
    cout<<"c3=";c3.display();
    c4=c2++;
    cout<<"c2=";c2.display();
    cout<<"c4=";c4.display();
    return 0;
}
```

程序运行结果如图 1.5.1 所示。

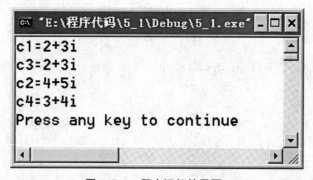

图 1.5.1　程序运行结果图

2. 声明一个复数类 Complex，类中设置两个私有数据 real 和 ima，分别表示复数的实部和虚部，类中声明成员函数 Complex operator＋(Complex c1)用来实现复数，友元函数 friend Complex operator－(Complex c1,Complex c2)用来实现复数对象减法运算。

复数类 Complex 声明如下：
```
#include <iostream.h>
class Complex
{
private:
    int real,imag;
```

```cpp
public:
    Complex(int r=0,int i=0);
    Complex operator+(Complex c1);
    friend Complex operator-(Complex c1,Complex c2);
    void display();
};
```

函数定义如下：

```cpp
Complex::Complex(int r,int i)
{
    real=r;
    imag=i;
}
Complex  Complex::operator+(Complex c1)
{
    Complex c;
    c.real=real+c1.real;
    c.imag=imag+c1.imag;
    return c;
}
Complex operator-(Complex c1,Complex c2)
{
    Complex c;
    c.real=c1.real-c2.real;
    c.imag=c1.imag-c2.imag;
    return c;
}
void Complex::display()
{
    cout<<real;
    if(imag>=0)
    {
        cout<<"+"<<imag<<"i";
    }
    else
    {
        cout<<imag<<"i";
    }
```

```
        cout<<endl;
}
```
主函数定义如下：
```
int main()
{
    Complex c1(1,2),c2(3,4),c3,c4;
    c3=c2+c1;
    cout<<"c2+c1=";
    c3.display();
    c4=c2-c1;
    cout<<"c2-c1=";
    c4.display();
    return 0;
}
```
程序运行结果如图 1.5.2 所示。

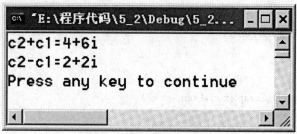

图 1.5.2　程序运行结果图

3. 首先应该声明一个实数矩阵类，该矩阵类的行数和列数是可变的，在类中设置两个变量 row 和 col 用来标识实际矩阵的行数和列数，设定矩阵的最大行数和列数均为 50。在类中声明一个矩阵元素赋值 evaluate(int,int,double)函数，以便用其对矩阵的元素赋初值。对于＋、－和 * 三种运算符的重载操作，通过使用友元函数 friend Matrix operator ＋(Matrix,Matrix)、friend Matrix operator －(Matrix,Matrix)和 friend Matrix operator * (Matrix,Matrix)来实现。对于()运算符用来返回矩阵中某元素的值得操作，采用成员函数 double operator()(int,int)来实现。

类 Matrix 的声明如下：
```
#include <iostream.h>
#include <iomanip.h>
const int Row=50;      //声明矩阵最大行数
const int Col=50;      //声明矩阵最大列数
class Matrix            //声明矩阵类
{
```

```cpp
private:
    int row,col;
    double matrix[Row][Col];
public:
    Matrix(int,int);
    double operator()(int,int);
    void evaluate(int,int,double);
    friend Matrix operator + (Matrix,Matrix);
    friend Matrix operator - (Matrix,Matrix);
    friend Matrix operator * (Matrix,Matrix);
    void print();
};
```

各函数的定义如下：

```cpp
Matrix::Matrix(int r,int c)                    //矩阵行数和列数的大小
{
    row=r;
    col=c;
}
double Matrix::operator ()(int r,int c)        //重载函数取矩阵r行c列的元素值
{
    return(r>=1&&r<=row&&c>=1&&c<=col)? matrix[r][c]:0.0;
}
void Matrix::evaluate(int r,int c,double value)    //设置矩阵r行c列的元素值
{
    if(r>=1&&r<=row&&c>=1&&c<=col)
    {
        matrix[r][c]=value;
    }
}
Matrix operator + (Matrix x,Matrix y)          //重载函数实现两个矩阵相加
{
    Matrix m(x.row,x.col);
    if(x.row!=y.row||x.col!=y.col){return m;}
    for(int r=1;r<=x.row;r++)
    {
        for(int c=1;c<=x.col;c++)
        {
```

```cpp
            m.evaluate(r,c,x(r,c)+y(r,c));
        }
    }
    return m;
}
Matrix operator-(Matrix x,Matrix y)              //重载函数实现两个矩阵相减
{
    Matrix m(x.row,x.col);
    if(x.row!=y.row||x.col!=y.col){return m;}
    for(int r=1;r<=x.row;r++)
    {
        for(int c=1;c<=x.col;c++)
        {
            m.evaluate(r,c,x(r,c)-y(r,c));
        }
    }
    return m;
}
Matrix operator*(Matrix x,Matrix y)              //重载函数实现两个矩阵相乘
{
    Matrix m(x.row,x.col);
    double temp;
    if(x.col!=y.row){ return m;}
    for(int r=1;r<=x.row;r++)
    {
        for(int c=1;c<=y.col;c++)
        {
            temp=0.0;
            for(int iter=1;iter<=x.col;iter++)
            {
                temp+=x(r,iter)*y(iter,c);
            }
            m.evaluate(r,c,temp);
        }
    }
    return m;
}
```

```
void Matrix::print()                          //输出矩阵
{
    for(int r=1;r<=this->row;r++)
    {
        for (int c=1;c<=this->col;c++)
        {
            cout<<setw(5)<<(*this)(r,c);
        }
        cout<<endl;
    }
}
```

主函数的定义如下：

```
int main()
{
    Matrix a(2,2),b(2,2),c(2,2),d(2,2),e(2,2);    //矩阵赋值
    a.evaluate(1,1,3.0);
    a.evaluate(1,2,5.0);
    a.evaluate(2,1,4.0);
    a.evaluate(2,2,6.0);
    b.evaluate(1,1,2.0);
    b.evaluate(1,2,0.9);
    b.evaluate(2,1,0.7);
    b.evaluate(2,2,0.1);                          //矩阵计算及结果输出
    cout<<setprecision(1)<<setiosflags(ios::fixed);
    cout<<"Matrix A:"<<endl;
    a.print();
    cout<<"Matrix B:"<<endl;
    b.print();
    c=a+b;
    d=a-b;
    e=a*b;
    cout<<"Matrix A+B:"<<endl;
    c.print();
    cout<<"Matrix A-B:"<<endl;
    d.print();
    cout<<"Matrix A*B:"<<endl;
    e.print();
```

```
        return 0;
    }
```

程序运行结果如图 1.5.3 所示。

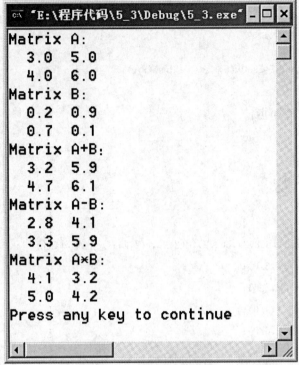

图 1.5.3　程序运行结果图

4. 声明一个复数类 Complex，类中设置两个私有数据 real 和 ima，分别表示复数的实部和虚部，友元函数 friend Complex operator＋(double d, Complex c)用来实现复数对象与 double 类型数据加法运算，即实现"d1＋c1"，类中声明成员函数 operator double( )用来实现复数将复数类型转换成 double 类型。

类 Complex 的声明如下：

```
#include <iostream.h>
class Complex
{
private:
    double real,imag;
public:
    Complex(int r=0,int i=0);
    friend Complex operator+(double d,Complex c);
    operator double( );
    void display();
```

};

各函数定义如下：

Complex::Complex(int r,int i)

{

  real=r;

  imag=i;

}

Complex operator+(double d,Complex c)

{

  Complex ct;

  ct.real=d+c.real;

  ct.imag=c.imag;

  return ct;

}

Complex::operator double( )

{

  return real;

}

主函数定义如下：

int main()

{

  double d1=10.3,d2;

  Complex c1(5,9);

  d2=d1+c1;

  cout<<d2<<endl;

  return 0;

}

程序运行结果如图 1.5.4 所示。

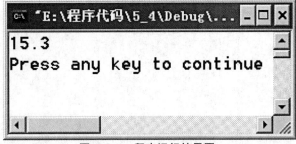

图 1.5.4　程序运行结果图

实验六　继承和派生(一)

一、实验目的

1. 理解继承和派生的概念,掌握派生类声明的方法,掌握派生类的构成。
2. 理解保护成员的访问属性。
3. 掌握公用继承、私有继承、保护继承和多级派生时,派生类各成员的访问属性。
4. 掌握派生类的构造函数和析构函数的声明规则。
5. 掌握简单派生类、有子对象派生类和多层派生时的构造函数和析构函数的调用顺序,特别是无参构造函数的调用过程。

二、实验内容

1. 有以下程序结构,请分析各成员的访问属性。

```
#include <iostream>
using namespace std;
class A                    //A为基类
{
public:
    void f1( );
    int i;
protected:
    void f2();
    int j;
private:
    int k;
};
class B: public A          //B为A的公用派生类
{
public:
    void f3( );
protected:
    int m;
private:
    int n;
```

```
};
class C：public B          //C 为 B 的公用派生类
{
public：
    void f4();
private：
    int p;
};
int main()
{
    A a1;              //a1 是基类 A 的对象
    B b1;              //b1 是派生类 B 的对象
    C c1;              //c1 是派生类 C 的对象
    return 0;
}
```

问题：

(1) 在 main 函数中能否用 b1.i,b1.j 和 b1.k 引用派生类 B 对象 b1 中基类 A 的成员？

(2) 派生类 B 中的成员函数能否调用基类 A 中的成员函数 f1 和 f2？

(3) 派生类 B 中的成员函数能否引用基类 A 中的数据成员 i,j,k？

(4) 能否在 main 函数中用 c1.i,c1.j,c1.k,c1.m,c1.n,c1.p 引用基类 A 的成员 i,j,k，派生类 B 的成员 m,n，以及派生类 C 的成员 p？

(5) 能否在 main 函数中用 c1.f1(),c1.f2(),c1.f3() 和 c1.f4() 调用 f1,f2,f3,f4 成员函数？

(6) 派生类 C 的成员函数 f4 能否调用基类 A 中的成员函数 f1,f2 和派生类 B 中的成员函数 f3？

2. 有以下程序，请完成下面工作。

(1) 阅读程序，写出运行时输出的结果；

(2) 上机运行，验证结果是否正确；

(3) 用单步调试分析程序执行过程，尤其是调用构造函数的过程。

```
#include <iostream>
using namespace std;
class A
{
public：
    A(){a=0;b=0;}
    A(int i){a=i;b=0;}
    A(int i,int j){a=i;b=j;}
```

```cpp
    void display(){cout<<"a="<<a<<" b="<<b;}
private:
    int a;
    int b;
};
class B : public A
{
public:
    B(){c=0;}
    B(int i):A(i){c=0;}
    B(int i,int j):A(i,j){c=0;}
    B(int i,int j,int k):A(i,j){c=k;}
    void display1()
    {
        display();
        cout<<" c="<<c<<endl;
    }
private:
    int c;
};
int main()
{
    B b1;
    B b2(1);
    B b3(1,3);
    B b4(1,3,5);
    b1.display1();
    b2.display1();
    b3.display1();
    b4.display1();
    return 0;
}
```

3. 有以下程序,请完成下面工作。

(1) 阅读程序,写出运行时输出的结果;

(2) 上机运行,验证结果是否正确;

(3) 用单步调试分析程序执行过程,尤其是调用构造函数的过程。

```cpp
#include <iostream>
using namespace std;
```

```cpp
class A
{
public：
    A(){cout<<"constructing A "<<endl;}
    ~A(){cout<<"destructing A "<<endl;}
};
class B  : public A
{
public：
    B(){cout<<"constructing B "<<endl;}
    ~B(){cout<<"destructing B "<<endl;}
};
class C  : public B
{
public：
    C(){cout<<"constructing C "<<endl;}
    ~C(){cout<<"destructing C "<<endl;}
};
int main()
{
    C c1;
    return 0;
}
```

4. 设计一个点 Point 类，从点 Point 类派生出圆 Circle 类，再从圆 Circle 类派生出圆柱 Cylinder 类，设计成员函数输出圆的面积以及圆柱的表面积和体积。

三、实验步骤

1. 根据题中给出继承关系，得出各类成员和访问属性如图 1.6.1 所示。

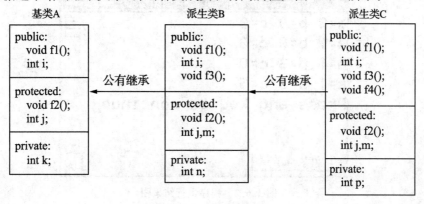

图 1.6.1 继承关系图

(1) B类是A类的公有派生类,i,j,k在B类的访问属性分别为公有、受保护和不可见,所在main函数中能用b1.i引用派生类B对象b1中基类A的成员,不能用b1.j和b1.k引用派生类B对象b1中基类A的成员。

(2) 在A类中,函数f1的访问属性为公有,所以派生类B中的成员函数能调用基类A中的成员函数f1;在A类中,函数f2的访问属性为受保护,所以派生类B中的成员函数能调用基类A中的成员函数f2。

(3) 在A类中,数据成员i的访问属性为公有,所以派生类B中的成员函数能引用基类A中的数据成员i;在A类中,数据成员j的访问属性为受保护,所以派生类B中的成员函数能引用基类A中的数据成员j;在A类中,数据成员k的访问属性为私有,所以派生类B中的成员函数不能引用基类A中的数据成员k。

(4) C类是B类的公有派生类,在C类中i的访问属性为公有,所以在main函数中可以用c1.i引用基类A的成员i;而j,m在C类中的访问属性为受保护,所以在main函数中不可以用c1.j,c1.m引用基类A的成员j和派生类B的成员m;k,n在C类中的访问属性为不可见的,所以在main函数中不可以用c1.k,c1.n引用基类A的成员k和派生类B的成员n;p是C类的私有成员,所以在main函数中不能用c1.p引用类C的成员p。

(5) f1,f3,f4在类C中的访问属性为公有,所以在main函数中能用c1.f1()、c1.f3()和c1.f4()调用f1,f3,f4成员函数;f2在类C中的访问属性为受保护的,所以在main函数中不能用c1.f2()调用f2成员函数。

(6) 在类B中,f1,f3的访问属性是公有的,所以派生类C的成员函数f4可以调用基类A中的成员函数f1和f3,在B类中,f2的访问属性是受保护的,所以派生类C的成员函数f4可以调用基类B中的成员函数f2。

2. 本题涉及构造函数的重载和简单派生类构造函数的调用顺序。简单派生类构造函数调用的顺序是,在调用派生类构造函数过程中先去调用基类构造函数,再执行派生类构造函数本身(即派生类构造函数的函数体)。程序运行结果图1.6.2所示。

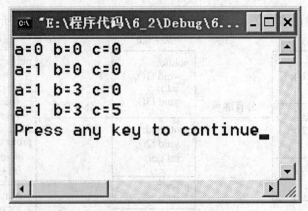

图 1.6.2 程序运行结果图

在执行语句"B b1;"时,调用派生类 B 的无参构造函数"B()",在调用"B()"时,首先去调用基类 A 的构造函数"A()"(当不需要向基类构造函数传递参数时,可以不写出基类构造函数的调用形式),执行完函数"A()"函数体后,再去执行函数"B()"的函数体,所以执行语句"b1.display1();"输出的结果如图 1.6.2 所示的第一行;在执行语句"B b2(1);"时,调用派生类 B 的一个参数的构造函数"B(int i):A(i)",在调用"B(int i):A(i)"时,首先去调用基类 A 的构造函数"A(int i)",执行完函数"A(int i)"函数体后,再去执行函数"B(int i):A(i)"的函数体,所以执行语句"b2.display1();"输出的结果如图 1.6.2 所示的第二行;在执行语句"B b3(1,3);"时,调用派生类 B 的两个参数的构造函数"B(int i,int j):A(i,j)",在调用"B(int i,int j):A(i,j)"时,首先去调用基类 A 的构造函数"A(int i,int j)",执行完函数"A(int i,int j)"函数体后,再去执行函数"B(int i,int j):A(i,j)"的函数体,所以执行语句"b3.display1();"输出的结果如图 1.6.2 所示的第三行;在执行语句"B b4(1,3,5);"时,调用派生类 B 的三个参数的构造函数"B(int i,int j,int k):A(i,j)",在调用"B(int i,int j,int k):A(i,j)"时,首先去调用基类 A 的构造函数"A(int i,int j)",执行完函数"A(int i,int j)"函数体后,再去执行函数"B(int i,int j,int k):A(i,j)"的函数体,所以执行语句"b4.display1();"输出的结果如图 1.6.2 所示的第四行。

3. 本题主要涉及简单派生类构造函数和析构函数调用顺序问题,在建立一个派生类对象时,在调用派生类构造函数过程中先去调用基类构造函数,再执行派生类构造函数本身(即派生类构造函数的函数体),在释放派生类对象时,先调用派生类析构函数,再调用基类析构函数。程序运行结果如图 1.6.3 所示。

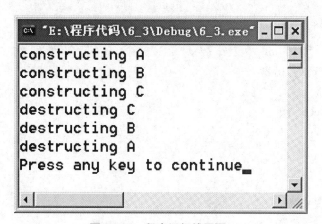

图 1.6.3 程序运行结果图

在执行语句"C c1;"时,调用 C 类的构造函数"C()",在调用构造函数"C()"过程中(没执行函数体前),去调用 B 类的构造函数"B()",在调用构造函数"B()"过程中(没执行函数体前),去调用 A 类的构造函数"A()"(执行函数体),所以输出结果如图 1.6.3 所示的第一行;A 类的构造函数调用完成后,返回到 B 类的构造函数处,执行 B 类构造函数的函数体,所以输出结果如图 1.6.3 所示的第二行;B 类的构造函数调用完成后,返回到 C 类的构造函数处，执行 C 类构造函数的函数体，所以输出结果如图 1.6.3 所示的第三行；当对象 c1 的生命周期结束时，要对对象 c1 所占用的存储空间做清理工作，也就是要调用析构函数，调用析构函数的顺序与调用构造函数的顺序是相反的，所以首先调用 C 类的析构函数"~C()"，输出结果如图 1.6.3 所示的第四行；其次调用 B 类的析构函数"~B()"，输出结果如图 1.6.3 所示的第五行；最后调用 A 类的析构函数"~A()"，输出结果如图 1.6.3 所示的第六行。

4. Point 类中有两个私有成员 x、y 用来表示直角坐标系中的点，Point 类的声明如下：

```
#include <iostream.h>
const float PI=3.14;
class Point
{
    int x,y;
};
```

以 Point 类为基类派生出派生类 Circle 类，在派生类中增加用来表示圆半径的受保护成员 r（以便在其派生类中被访问），同时增加 Circle 类的构造函数、求面积的 area 函数和输出面积的 print 函数。Circle 类的声明如下：

```
class Circle:public Point
{
protected:
    int r;
public:
    Circle(int r=0);
    double area();
    void print();
};
Circle::Circle(int r)
{
    this->r=r;
}
double Circle::area()
{
    return PI*r*r;
}
void Circle::print()
{
    cout<<"圆的面积:"<<area()<<endl;
}
```

以 Circle 类为直接基类派生出派生类 Cylinder 类,在派生类中增加用来表示圆柱高的私有成员 h,同时增加 Cylinder 类的构造函数、计算表面积的 area 函数、计算体积的 volume 函数和输出面积、体积的 print 函数。Cylinder 类的声明如下：

```cpp
class Cylinder:public Circle
{
    int h;
public:
    Cylinder(int r,int h);
    double area();
    double volume();
    void print();
};
Cylinder::Cylinder(int r,int h):Circle(r)
{
    this->h=h;
}
double Cylinder::area()
{
    return 2*PI*r*r+2*PI*r*h;
}
double Cylinder::volume()
{
    return PI*r*r*h;
}
void Cylinder::print()
{
    cout<<"圆柱的表面积是:"<<area()<<endl;
    cout<<"圆柱的体积是:"<<volume()<<endl;
}
```

主函数的定义如下：

```cpp
int main( )
{
    Circle cir(2);
    Cylinder cyl(1,2);
    cir.print();
    cyl.print();
    return 0;
}
```

程序运行结果如图 1.6.4 所示。

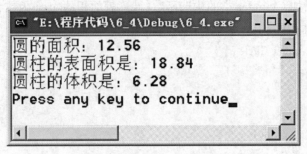

图 1.6.4　程序运行结果图

实验七 继承和派生(二)

一、实验目的

1. 进一步理解继承和派生的概念,掌握派生类构造函数的书写方法,理解派生类构造函数和析构函数执行顺序。

2. 了解多重继承的含义,掌握多重继承时引起二义性问题的解决方法,理解虚基类的含义以及使用方法。

3. 掌握基类与派生类的转换,即派生类对象可以向基类对象赋值、派生类对象可以替代基类对象向基类对象的引用进行赋值或初始化和派生类对象可以赋给指向基类对象的指针变量。

4. 了解继承与组合的含义。

5. 理解继承在软件开发中的重要意义。

二、实验内容

1. 设计一个点类 Point,再设计一个矩形 Rectangle 类,矩形 Rectangle 类使用 Point 类的两个坐标点作为矩形的对角顶点,并可以输出 4 个坐标值和面积。使用测试程序验证程序。

2. 动物中包含两大类:脊椎动物和无脊椎动物。脊椎动物中再细分有哺乳动物、鸟类和其他类别。从脊椎动物到哺乳动物和鸟类再到具体的一种动物,每一层都有共性的信息,也有其本身个性的信息。试编写一个统计动物物种种类的程序,对每一种具体的动物都能给出其物种的共性特征,也能给出其本身的个性信息。

三、实验分析

1. 首先,在 Point 类中有设置两个私有成员 x、y 用来表示直角坐标系中的点,成员函数有用来初始化私有成员数据的构造函数 Point,设置坐标的 setXY 函数,得到坐标 x,y 的 getX,getY 函数,Point 类的声明如下:

```
#include <iostream>
using namespace std;
class Point                //点类
{
private:
    int x, y;//私有成员变量,坐标
```

public:

  Point();//无参数的构造方法,对 xy 初始化

  Point(int a, int b); //有参数的构造方法,对 xy 初始化

  void setXY(int a, int b);//设置 xy 的值

  int getX();//得到 x 的方法

  int getY();//得到 y 的函数

};

Point::Point()

{

  x = 0;

  y = 0;

}

Point::Point(int a, int b)

{

  x = a;

  y = b;

}

void Point::setXY(int a, int b)

{

  x = a;

  y = b;

}

int Point::getX()

{

  return x;

}

int Point::getY()

{

  return y;

}

其次,在 Rectangle 类中定义 point1,point2,point3,point4 四个私有成员用来表示矩形的四个点;三个构造函数分别是:无参构造函数 Rectangle(),用两个 Point 类对象作为形参的构造函数 Rectangle(Point one, Point two),用 4 个坐标值作为函数参数的构造函数 Rectangle(int x1, int y1, int x2, int y2);给矩形另外两个顶点初始化的函数 init(),输出矩形四个顶点的 printPoint() 输出函数,计算矩形面积的 getArea() 函数。

  class Rectangle   //矩形类

  {

```cpp
private:
    Point point1, point2, point3, point4;
public:
    Rectangle();        //类 Point 的无参构造函数已经对每个对象做初始化，
                        //这里不用对每个点多初始化
    Rectangle(Point one, Point two);//用 Point 类对象作为构造函数的 2 个参数
    Rectangle(int x1, int y1, int x2, int y2);//用 4 个坐标值作为构造函数的 4 个参数
    void init();//给另外两个点做初始化的函数
    void printPoint();//打印四个点的函数
    int getArea();//计算面积的函数
};
Rectangle::Rectangle(Point one, Point two)
{
    point1 = one;
    point4 = two;
    init();
}
Rectangle::Rectangle(int x1, int y1, int x2, int y2)
{
    point1.setXY(x1, y1);
    point4.setXY(x2, y2);
    init();
}
void Rectangle::init()
{
    point2.setXY(point4.getX(), point1.getY());
    point3.setXY(point1.getX(), point4.getY());
}
void Rectangle::printPoint()
{
    cout<<"A:("<< point1.getX() <<","<< point1.getY() <<")"<< endl;
    cout<<"B:("<< point2.getX() <<","<< point2.getY() <<")"<< endl;
    cout<<"C:("<< point3.getX() <<","<< point3.getY() <<")"<< endl;
    cout<<"D:("<< point4.getX() <<","<< point4.getY() <<")"<< endl;
}
int Rectangle::getArea()
{
```

```
    int height, width, area;
    height = point1.getY() - point3.getY();
    width = point1.getX() - point2.getX();
    area = height * width;
    if(area > 0)
        return area;
else
        return -area;
}
```

最后,定义主函数,主函数代码如下:

```
void main()
{
    Point p1(-15, 56), p2(89, -10);//定义两个点
    Rectangle r1(p1, p2);//用两个点做参数,声明一个矩形对象r1
    Rectangle r2(1, 5, 5, 1);//用四个整数做参数,声明一个矩形对象r2
    cout<<"矩形 r1 的 4 个定点坐标:"<< endl;
    r1.printPoint();
    cout<<"矩形 r1 的面积:"<< r1.getArea() << endl;
    cout<<"\n 矩形 r2 的 4 个定点坐标:"<< endl;
    r2.printPoint();
    cout<<"矩形 r2 的面积:"<< r2.getArea() << endl;
}
```

程序运行结果如图 1.7.1 所示。

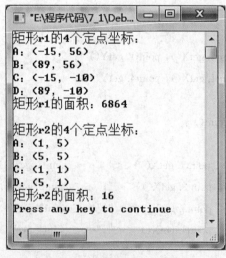

图 1.7.1　程序运行结果图

2. 根据题目的要求,首先建立一个建筑类 Architecture,该类中包含建筑物的基本信

息,数据成员有建筑物的高度 height、占地面积 area_a 和建筑物类型 type,构造函数 Architecture 用来初始化建筑物对象,成员函数 ShowInfo 用于输出建筑物的高度、占地面积和类型。具体代码如下:

```cpp
#include <iostream>
using namespace std;
class Architecture
{
public:
    Architecture(double h,double a,int );
    void ShowInfo();
    void ArchType(int t);
protected:
    double height,area_a;
    int type;
};
Architecture::Architecture (double h=0,double a=0,int t=0)
{
    height=h;
    area_a=a;
    type=t;
}
void Architecture::ArchType(int t)
{
    if(t==0)
    {cout<<"楼房"<<endl;}
    if(t==1)
    {cout<<"楼房"<<endl;}
    if(t==2)
    {cout<<"楼房"<<endl;}
}
void Architecture::ShowInfo()
{
    cout<<endl;
    cout<<"建筑物高度:"<<height<<"米"<<endl;
    cout<<"建筑物占地面积:"<<area_a<<"平方米"<<endl;
    cout<<"建筑物类型:";
    ArchType(type);
```

其次，以建筑物类 Architecture 为基类公有派生出楼房类 Building，新增加了保护数据成员 floors 用于表示楼房的层数、rooms 用于表示楼房的房间数和 area_b 用于表示楼房的总面积，新增加了公有构造函数 Building 用来初始化楼房对象和 ShowInfo 函数用于输出楼房的层数 floors、房间数 rooms 和总面积 area_b。具体代码如下：

```cpp
class Building:public Architecture
{
public:
    Building(double h=0,double a=0,int t=0,int f=0,int r=0,double b=0);
    void ShowInfo();
protected:
    int floors,rooms;
    double area_b;
};
Building::Building(double h,double a,int t,int f,int r,double b):Architecture(h,a,t)
{
    floors=f;
    rooms=r;
    area_b=b;
}
void Building::ShowInfo()
{
    Architecture::ShowInfo();
    cout<<"楼房层数:"<<floors<<"层"<<endl;
    cout<<"楼房房间数:"<<rooms<<"间"<<endl;
    cout<<"楼房总面积:"<<area_b<<"平方米"<<endl;
    cout<<"其中:";
}
```

第三，以楼房类 Building 为基类公有派生出住宅类 Home，新增加了私有数据成员卧室数 bedrooms、卫生间数 toilets 和厨房数 kitchens，新增加了公有构造函数 Home 用来初始化住宅类的对象和 show 函数用于输出住宅类的卧室数 bedrooms、卫生间数 toilets 和厨房数 kitchens。具体代码如下：

```cpp
class Home:public Building
{
public:
    Home(double h=0,double a=0,int t=0,int f=0,int r=0,
```

```
        double b=0,int br=0,int to=1,int k=1);
        void show();
    private:
        int bedrooms,toilets,kitchens;
};
Home::Home(double h,double a,int t,int f,int r,double b,int br,int to,int k):Building(h,a,t,f,r,b)
{
    bedrooms=br;
    toilets=to;
    kitchens=k;
}
void Home::show()
{
    cout<<endl;
    cout<<"住宅数:";
    Building::ShowInfo();
    cout<<"卧室数:"<<bedrooms<<"间"<<endl;
    cout<<"       卫生间数:"<<toilets<<"间"<<endl;
    cout<<"       厨房数:"<<kitchens<<"间"<<endl;
}
```

第四,以楼房类 Building 为基类公有派生出医院类 Hospital,新增加了私有数据成员病房数 sickrooms 和手术室数 operating_rooms,新增加了公有构造函数 Hospital 用来初始化医院的对象和 show 函数用于输出医院类的病房数 sickrooms 和手术室数 operating_rooms。具体代码如下:

```
class Hospital:public Building
{
public:
    Hospital(double h=0,double a=0,int t=0,int f=0,int r=0,int b=0,int s=0,int or=0);
    void show();
private:
    int sickrooms,operating_rooms;
};
Hospital::Hospital(double h,double a,int t,int f,int r,int b,int s,int or):Building(h,a,t,f,r,b)
{
    sickrooms=s;
    operating_rooms=or;
```

}
void Hospital::show()
{
    cout<<endl;
    cout<<"医院";
    Building::ShowInfo();
    cout<<"病房数:"<<sickrooms<<"间"<<endl;
    cout<<"   手术室数:"<<operating_rooms<<"间"<<endl;
}

第五，以楼房类 Building 为基类公有派生出办公楼类 Office,新增加了私有数据成员电话数 phones 和灭火器数 fire,新增加了公有构造函数 Office 用来初始化办公楼的对象和 show 函数用于输出办公楼类的电话数 phones 和灭火器数 fire。具体代码如下：

```
class Office:public Building
{
public:
    Office(double h=0,double a=0,int t=0,int f=0,int r=0,int b=0,int p=0,int fe=0);
    void show();
private:
    int phones,fire;
};
Office::Office(double h,double a,int t,int f,int r,int b,int p,int fe):Building(h,a,t,f,r,b)
{
    phones=p;
    fire=fe;
}
void Office::show()
{
    cout<<endl;
    cout<<"办公楼";
    Building::ShowInfo();
    cout<<"电话数:"<<phones<<"部"<<endl;
    cout<<"      灭火器数:"<<fire<<"个"<<endl;
}
```

最后，定义用于测试类的主函数,主函数代码如下：

```
int main()
{
    Home home(25,500,0,7,100,14000,2,2,1);
```

Hospital hospital(45,1000,0,10,600,45000,400,50);

Office office(16,700,0,4,40,3500,40,12);

home.show();

hospital.show();

office.show();

return 0;

}

程序运行结果如图 1.7.2 所示。

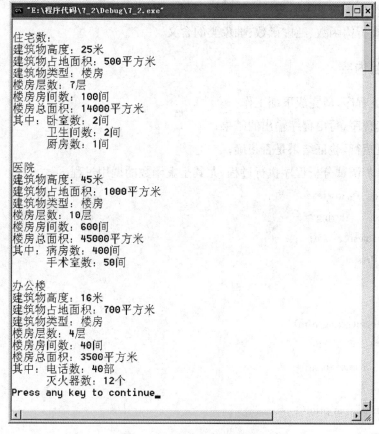

图 1.7.2　程序运行结果图

实验八 多态性与虚函数

一、实验目的

1. 理解多态的概念,理解静态多态与动态多态的区别。
2. 理解虚函数的含义,掌握虚函数的声明方法,理解虚函数的应用场合。
3. 掌握一个函数被声明为虚函数前后调用过程的区别。
4. 了解虚析构函数、纯虚函数、抽象类的含义。

二、实验内容

1. 有以下程序,请完成下面工作。
(1) 阅读程序,写出程序输出的结果;
(2) 上机运行,验证结果是否正确;
(3) 用单步调试分析程序执行过程,尤其是虚函数的调用过程。

```cpp
#include <iostream>
#include <string>
using namespace std;
class Person
{
public:
    Person(string nam)
    {
        name=nam;
    }
    virtual void print()
    {
        cout<<"我的名字是"<<name<<"。\n";
    }
protected:
    string name;
};
class Student:public Person
{
public:
```

```cpp
        Student(string nam,float g):Person(nam)
        {
            ga=g;
        }
        void print( )
        {
            cout<<s<<"我的名字是"<<name<<",我的年薪是"<<ga<<"万元。\n";
        }
    private:
        float ga;
};
class Professor:public Person
{
public:
        Professor(string nam,int n):Person(nam)
        {
            publs=n;
        }
        void print( )
        {
            cout<<"我的名字是"<<name<<",我已出版了"<<publs<<"本教材。\n";
        }
    private:
        int publs;
};

void main( )
{
    Person *p;
    Person x("张刚");
    Student y("王强",4.2);
    Professor z("李明",5);
    p=&x; p->print( );
    p=&y; p->print( );
    p=&z; p->print( );
}
```

2. 在几何学中,先是有点的概念,然后才是由点构成的其他各种几何形状。例如,在平面中到定点的距离都相等的所有的点构成了一个圆,由圆又可以推出圆柱体的概念。点没有体积和面积的概念,圆有面积的概念而无体积的概念,而圆柱体兼有体积和表面积两者的概念。试使用C++中类的概念编程实现上述关系,输出各种几何形状的表象。

3. 某软件公司现有三类人员:行政管理人员、项目开发人员和项目开发部门的管理人员(既承担行政管理工作,又参与项目开发)。现在需要对公司的人员信息进行统一管理,存储人员的编号、姓名、职务级别、固定月薪和计算每月奖金,并且能够显示其全部月收入。

人员的编号基数为8000,每当新增加一个人员时编号顺序加1即可。

行政管理人员和项目开发人员均划分成三个等级,行政管理人员分为总经理、部门经理和小组长;项目开发人员分为工作时间不满一年的、工作时间超过一年不到三年的和工龄在三年以上的。

行政管理人员的最高级别为总经理,每月工资为12000元,固定月薪的计算公式为12000*(3－级别＋1)/3;每月固定奖金为3500元。

项目开发人员的最高级别是工龄在三年以上的员工,每月工资为6000元,固定月薪的计算公式为6000*(3－级别＋1)/3＋奖金为其加班的小时数*40元/小时＋500元。

项目开发部门的管理人员的待遇同项目开发人员的待遇,另外还有再加上小组长级别的固定月薪。试编程实现上述人员管理。

三、实验分析

1. 程序运行结果如图1.8.1所示,该题定义了一个基类Person,以Person为基类派生出派生类Student和Professor,在基类定义一个虚函数print,在两个派生类中分别重写了print函数。在主函数中定义一个指向基类的指针变量p,定义了三个类Person、Student和Professor的对象x、y和z,当执行语句"p=&x;"时,p指向基类对象x,执行语句时"p→print( );",调用基类的print函数,所以输出结果如图1.8.1所示的第一行;当执行语句"p=&y;"时,p指向派生类对象y,执行语句时"p→print( );",调用派生类Student的print函数,所以输出结果如图1.8.1所示的第二行;当执行语句"p=&z;"时,p指向派生类对象z,语句"p→print( );",调用派生类Professor的print函数,所以输出结果如图1.8.1所示的第三行。

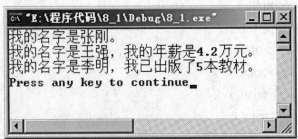

图1.8.1 程序运行结果图

2. 从点→圆→圆柱体,利用类的继承和派生,可以很容易地实现这种转换关系。考虑到所有的几何形状,先声明一个最底层的抽象基类 From(形状),这样所有的其他几何图形的类都是这个基类的直接或间接派生类,在 From 类中声明三个成员函数,分别是面积、体积和几何形状名称,三个成员函数都声明为虚函数。利用 From 类派生出点的概念,进而获得圆和圆柱体的类。对于点、面积和体积都设为 0,圆柱的体积设为 0。

程序的实现流程可描述如下:

(1) 声明一个抽象形状基类 From,代码如下:

```
#include <iostream.h>
const float PI=3.14;
class Form        //声明形状基类
{
public:
    virtual double area()    {return 0.0;}  //声明面积虚函数
    virtual double bulk()    {return 0.0;}  //声明体积虚函数
    virtual void formName()   =0;           //声明形状纯虚函数
};
```

(2) 由 From 类派生出点 Point 类,代码如下:

```
class Point:public Form    //声明点类
{
public:
    Point(double a=0,double b=0);
    void setPoint(double a,double b);
    double getX()    {return x;}   //声明成员函数返回点坐标
    double getY()    {return y;}
    virtual void formName()    {cout<<"点:";}  //声明类的形状类型
    friend ostream & operator<<(ostream &, Point &);
protected:
    double x;
    double y;
};
Point::Point(double a,double b)
{
    x=a;
    y=b;
}
void Point::setPoint(double a,double b)
```

```
    {
        x=a;
        y=b;
    }
    ostream & operator<<(ostream &output, Point &p)
    {
        output<<"("<<p.x<<","<<p.y<<")";
        return output;
    }
```

(3) 由 Point 类派生出圆 Circle 类,代码如下:
```
class Circle:public Point                    //声明圆类
{
public:
    Circle(double x=0,double y=0,double r=0);
    void setRadius(double);
    double getRadius()   ;                   //声明返回圆的半径的成员函数
    virtual double girth()   ;               //声明返回圆的周长的成员函数
    virtual double area()   ;                //声明返回圆的面积的成员函数
    virtual void formName()                  //声明返回圆的形状的成员函数
    {cout<<"圆:";}
    friend ostream & operator<<(ostream &, Circle &);
protected:
    double radius;
};
Circle::Circle(double x,double y,double r):Point(x,y),radius(r){}
void Circle::setRadius(double r)
{
    radius=r;
}
double Circle::getRadius()
{
    return radius;
}
double Circle::girth()
{
    return 2*PI*radius;
}
```

```cpp
double Circle::area()
{
    return PI * radius * radius;
}
ostream & operator<<(ostream &output, Circle &c)
{
    output<<"("<<c.x<<","<<c.y<<","<<"),r="<<c.radius;
    return output;
}
```

(4) 由 Circle 类派生出圆柱体 Cylinder 类,代码如下:

```cpp
class Cylinder:public Circle
{
public:
    Cylinder(double a=0,double b=0,double r=0,double h=0);
    void setHeight(double);              //声明返回圆柱体的高的成员函数
    virtual double area()    ;           //声明返回圆柱体的表面积的成员函数
    virtual double bulk()    ;           //声明返回圆柱体的体积的成员函数
    virtual void formName()              //声明返回圆柱体的形状的成员函数
    {
        cout<<"圆柱体:";
    }
    friend ostream & operator<<(ostream &, Cylinder &);
protected:
    double height;
};
Cylinder::Cylinder(double a,double b,double r,double h):
    Circle(a,b,r),height(h)
    {}
void Cylinder::setHeight(double h)
{
    height=h;
}
double Cylinder::area()
{
    return 2 * Circle::area()+Circle::girth() * height;
}
double Cylinder::bulk()
```

```
    {
        return Circle::area() * height;
    }
    ostream & operator<<(ostream & output, Cylinder & cy)
    {
        output<<"("<<cy.x<<","<<cy.y<<","<<")"<<",r="<<cy.radius<<",h="<<cy.height;
        return output;
    }
```

(5) 建立类的对象进行验证,代码如下:

```
    int main()
    {
        Point p(5.5,5.8);
        Circle c(5.5,5.8,12.0);
        Cylinder cy(5.5,5.8,12.0,31.2);
        Form * pt;
        pt=&p;
        pt->formName();
        cout<<p<<endl;
        cout<<"面积="<<pt->area()<<"\n 体积="<<pt->bulk()<<endl;
        cout<<endl;
        pt=&c;
        pt->formName();
        cout<<"面积="<<pt->area()<<"\n 体积="<<pt->bulk()<<endl;
        cout<<endl;
        pt=&cy;
        pt->formName();
        cout<<cy<<endl;
        cout<<"面积="<<pt->area()<<"\n 体积="<<pt->bulk()<<endl;
        cout<<endl;
        return 0;
    }
```

程序运行结果如图 1.8.2 所示。

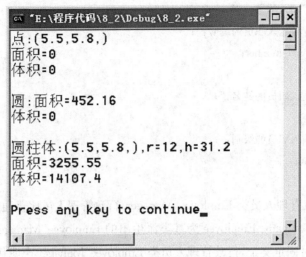

图 1.8.2　程序运行结果图

3. 对于三类公司职员，都有其共同的特征：一个雇员的基本信息。一个雇员的基本信息包括这个雇员的编号、姓名、职务级别和月收入，所以首先声明一个公司雇员的抽象基类 Company_Employee，包含一个基本雇员的信息，Company_Employee 类声明代码如下：

```
#include <iostream>
#include <string>
using namespace std;
class Company_Employee                //定义公司雇员的基类
{
public：
    Company_Employee()；
    virtual void Month_Pay()=0;       //计算月薪虚函数
    virtual void Show_Status()=0;     //显示雇员信息虚函数
protected：
    string name;                      //姓名
    int serial_number;                //编号
    int position;                     //职务级别
    float monthly_pay;                //月收入
};
Company_Employee：：Company_Employee()
{
cout<<"请输入职员的编号：";
cin>>serial_number;
while(serial_number<8000)
```

```
        {
            cout<<"输入编号错误,请重新输入!"<<endl;
            cout<<"请输入职员的编号:";
            cin>>serial_number;
        }
        cout<<"请输入职员的姓名:";
        cin>>name;
        cout<<"请输入职员的级别:";
        cin>>position;
    }
```

其次,声明行政管理人员类 Employee_Manager,行政管理人员比普通的雇员增加了每月的固定奖金,所以以 Company_Employee 为基类派生出的 Employee_Manager 派生类中增加成员 month_pay 表示每月固定奖金,行政管理人员类 Employee_Manager 声明代码如下:

```
class Employee_Manager:virtual public Company_Employee    //行政管理人员类
{
public:
    Employee_Manager();
    void Month_Pay();
    void Show_Status();
private:
    int month_pay;//每月奖金
};
Employee_Manager::Employee_Manager()
{
    month_pay=3500;
}
void Employee_Manager::Month_Pay()
{
    monthly_pay=12000*(3-position+1)/3+month_pay;
    cout<<"行政管理人员"<<name<<"月收入为:"<<monthly_pay<<"元"<<endl;
}
void Employee_Manager::Show_Status()
{
    cout<<"行政管理人员"<<name<<"编号"<<serial_number<<"级别"<<position<<"本月收入"<<monthly_pay<<"元"<<endl;
    cout<<endl;
}
```

第三,声明项目开发人员类 Employee_Developer,项目开发人员比普通的雇员增加了加班的小时数和每月的固定奖金,所以以 Company_Employee 为基类派生出的 Employee_Developer 派生类中增加成员 hour_salary 表示每小时的加班费、work_hours 表示加班累计小时数和 month_pay 表示每月固定奖金,所以项目开发人员类 Employee_Developer 声明代码如下:

```cpp
class Employee_Developer:virtual public Company_Employee     //项目开发人员类
{
public:
    Employee_Developer();
    void Month_Pay();
    void Show_Status();
protected:
    int hour_salary;                    //每小时的加班费
    int work_hours;                     //加班累计小时数
    int month_pay;                      //月奖金额
};
Employee_Developer::Employee_Developer()
{
    cout<<"请输入职员的加班累计小时数:";
    cin>>work_hours;
    hour_salary=40;
    month_pay=500;
}
void Employee_Developer::Month_Pay()
{
    monthly_pay=6000*(3-position+1)/3+work_hours*hour_salary+500;
    cout<<"项目开发人员"<<name<<"本月收入"<<monthly_pay<<"元"<<endl;
}
void Employee_Developer::Show_Status()
{
    cout<<"项目开发人员"<<name<<"编号"<<serial_number<<"级别"<<position<<"本月收入"<<monthly_pay<<"元"<<endl;
    cout<<endl;
}
```

第四,声明项目开发部门管理人员类 Employee_Branch,项目开发部门管理人员同时具有行政管理人员和项目开发人员的特点,所以由行政管理人员类 Employee_Manager 和项目开发人员类 Employee_Developer 派生出 Employee_Branch，因为行政管理人员类 Employee_Manager 和项目开发人员类 Employee_Developer 含有相同的成员，为了让它们相同的成员在项目开发部门管理人员类 Employee_Branch 只有一个拷贝，所以在声明行政管理人员类 Employee_Manager 和项目开发人员类 Employee_Developer 时，将它们声明为虚基类。项目开发部门管理人员类 Employee_Branch 类的声明代码如下：

```cpp
class Employee_Branch:public Employee_Developer,public Employee_Manager//项目管理人员类
{
public：
    Employee_Branch() {}
    void Month_Pay();
    void Show_Status();
};
void Employee_Branch::Month_Pay()
{
    monthly_pay=6000*(3-position+1)/3+work_hours*hour_salary+12000*(3-3+1)/3
+Employee_Developer::month_pay;
    cout<<"项目管理人员"<<name<<"月收入为："<<monthly_pay<<"元"<<endl;
}
void Employee_Branch::Show_Status()
{
    cout<<"项目管理人员"<<name<<"编号"<<serial_number<<"级别"<<position<<"本月收入"<<monthly_pay<<"元"<<endl;
    cout<<endl;
}
```

第五，定义用来管理菜单的 Menu_Manager 类，该类主要实现菜单的显示和接受菜单选择的功能，Menu_Manager 类声明代码如下：

```cpp
class Menu_Manager             //菜单类
{
public：
    Menu_Manager() {}
    void Menu_Choice();        //显示菜单选项函数
    char Get_Choice();         //接受菜单选择函数
};
void Menu_Manager::Menu_Choice()
{
    cout<<"请选择菜单："<<endl;
    cout<<"        行政管理(M 或 m)"<<endl;
```

```cpp
        cout<<"        开发人员(D 或 d)"<<endl;
        cout<<"        项目管理(B 或 b)"<<endl;
        cout<<"        退出系统(Q 或 q)"<<endl;
        cout<<"请输入你的选择:";
}
char Menu_Manager::Get_Choice()
{
        char ch;
        cin>>ch;
        return ch;
}
```

最后,定义主函数,主函数代码如下:

```cpp
void main()
{
        char ch;
        Menu_Manager *menu=new Menu_Manager;
        menu->Menu_Choice();
        ch=menu->Get_Choice();
        while(ch!='q'&&ch!='Q')            //循环查询雇员信息
        {
                switch(ch)
                {
                case'd':
                case'D':
                {
                        Employee_Developer *developer=new Employee_Developer;
                        developer->Month_Pay();
                        developer->Show_Status();
                        delete developer;
                        break;
                }
                case'm':
                case'M':
                {
                        Employee_Manager *manager=new Employee_Manager;
                        manager->Month_Pay();
                        manager->Show_Status();
```

```
                delete manager;
                break;
            }
        case 'b':
        case 'B':
            {
                Employee_Branch * Branch=new Employee_Branch;
                Branch->Month_Pay();
                Branch->Show_Status();
                delete Branch;
                break;
            }
        default:
            {
                cout<<"菜单选择错误,请重新选择!"<<endl;
                break;
            }
        }
        menu->Menu_Choice();
        ch=menu->Get_Choice();
    }
    delete menu;
}
```

程序运行结果如图 1.8.3 所示。

```
"E:\程序代码\8_3\Debug\8_3.exe"
请选择菜单：
        行政管理   (M或m)
        开发人员   (D或d)
        项目管理   (B或b)
        退出系统   (Q或q)
请输入你的选择：m
请输入职员的编号：80001
请输入职员的姓名：zhangsan
请输入职员的级别：2
行政管理人员zhangsan月收入为：11500元
行政管理人员zhangsan编号80001级别2本月收入11500元

请选择菜单：
        行政管理   (M或m)
        开发人员   (D或d)
        项目管理   (B或b)
        退出系统   (Q或q)
请输入你的选择：d
请输入职员的编号：8002
请输入职员的姓名：lisi
请输入职员的级别：1
请输入职员的加班累计小时数：50
项目开发人员lisi本月收入8500元
项目开发人员lisi编号8002级别1本月收入8500元

请选择菜单：
        行政管理   (M或m)
        开发人员   (D或d)
        项目管理   (B或b)
        退出系统   (Q或q)
请输入你的选择：b
请输入职员的编号：8003
请输入职员的姓名：wangwu
请输入职员的级别：2
请输入职员的加班累计小时数：50
项目管理人员wangwu月收入为：10500元
项目管理人员wangwu编号8003级别2本月收入10500元

请选择菜单：
        行政管理   (M或m)
        开发人员   (D或d)
        项目管理   (B或b)
        退出系统   (Q或q)
请输入你的选择：q
Press any key to continue
```

图 1.8.3 程序运行结果图

实验九 输入输出流

一、实验目的

1. 理解C++的输入输出的含义。
2. 理解C++标准输出流和标准输入流,掌握标准输出流对象 cout、cerr 和 clog 的使用,掌握标准输入流对象 cin 的使用。
3. 理解C++中文件的概念。
4. 理解文件流类和文件流对象的概念。
5. 掌握C++中对 ACSII 码文件和二进制文件的操作。
6. 理解字符串流的含义。

二、实验内容

1. 输出杨辉三角,使用流的格式化状态字与 I/O 操作算子实现输出的左对齐、居中对齐和右对齐三种对齐方式。
2. 编写程序,将一个文本文件中的所有数字字符读出并拷贝到另一个文本文件中去。
3. 现有若干名学生的成绩数据,要求编程完成如下功能:
(1) 把学生的成绩信息存入到磁盘文件中;
(2) 从磁盘文件中读出某一条学生的信息,并显示出来;
(3) 将某条学生信息修改后存回磁盘文件中的原有位置;
(4) 将修改后的学生信息读入内存并显示出来;
(5) 磁盘文件采用二进制数据文件。

三、实验分析

1. 所谓杨辉三角,其数值就是对应于形如(a+b)^n 的 n 次多项式的展开式的各项系数。它具有如下规律:当排成如下形式时,它的第一层的数是1,其余各层的最左侧和最右侧的数也均是1,其余各层中其他的数分别是其对应两个"肩膀"上的数的和。

```
        1
       1 1
      1 2 1
     1 3 3 1
    1 4 6 4 1
      ……
```

用 m 表示层数,a[m][n]表示 m 层的第 n 个元素,则有:

① 当 n=0 或 n=m 时,有 a[m][n]=1;
② 当 0<n<m 时,有 a[m][n]=a[m-1][n-1]+a[m-1][n]。
程序的实现流程描述如下:
(1) 声明关于杨辉三角的类,该类具有带参数的构造函数、析构函数和实现三种输出方式的成员函数;
(2) 选择输出的方式;
(3) 按照该方式输出杨辉三角。

```cpp
#include <iostream.h>
#include <iomanip.h>
const int MAXLINE=20;
class YangHui                    //定义杨辉三角的类
{
public:
    YangHui(int);                //声明构造函数
    ~YangHui();                  //声明析构函数
    void lalign_out();           //声明输出方式的函数
    void ralign_out();
    void malign_out();
private:
    int * y[MAXLINE];
    int n;
};
YangHui::YangHui(int n)          //定义构造函数
{
    YangHui::n=n;
    for(int i=0;i<n;i++)
    {
        y[i]=new int[sizeof(int)*(i+1)];
        for(int j=0;j<=i;j++)
        {
            if(j==0||j==i)
            {
                *(y[i]+j)=1;
            }
            else
            {
                *(y[i]+j)= *(y[i-1]+j-1)+ *(y[i-1]+j);
            }
```

```cpp
        }
    }
}
void YangHui::lalign_out()            //定义左对齐输出方式
{
    cout<<setiosflags(ios::left);   //使用流的格式化状态字确定数据左对齐
    for(int i=0;i<n;i++)
    {
        for(int j=0;j<=i;j++)
        {
            cout<<setw(5);
            cout<<*(y[i]+j);
        }
        cout<<endl;
    }
}
void YangHui::ralign_out()            //定义右对齐输出方式
{
    cout<<setiosflags(ios::right);  //使用流的I/O操作算子确定数据右对齐
    cout<<endl;
    for(int i=0;i<n;i++)
    {
        for(int j=0;j<=i;j++)
        {
            if(j==0)
            {
                cout<<setw((n-i)*5-j*n);
            }
            else
            {
                cout<<setw(5);
            }
            cout<<*(y[i]+j);
        }
        cout<<endl;
    }
}
void YangHui::malign_out()            //定义居中对齐输出方式
{
```

```cpp
        cout<<endl;
        for(int i=0;i<n;i++)
        {
            for(int j=0;j<=i;j++)
            {
                if(j==0)
                {
                    cout<<setw(((n-i)*5-j*n)/2)<<*(y[i]+j);
                }
                else
                {
                    cout<<setw(5)<<*(y[i]+j);
                }
            ;
            }
            cout<<endl;
        }
    }
    YangHui::~YangHui()              //定义析构函数
    {
        for(int i=0;i<n;i++)
        {
            delete y[i];
        }
    }
    int main()
    {
        YangHui trianhle(9);         //声明一个类的对象
        char lmr,con;                //选择输出对齐方式
        do
        {
            cout<<"输出左对齐、居中对齐还是右对齐杨辉三角(L/M/R):";
            cin>>lmr;                //按要求的方式输出杨辉三角
            if(lmr=='l'||lmr=='L')
            {
                trianhle.lalign_out();
            }
            else if(lmr=='r'||lmr=='R')
            {
```

```
                trianhle.ralign_out();
            }
            else if(lmr=='m'||lmr=='M')
            {
                trianhle.malign_out();
            }
            cout<<"是否继续？(Y/N)";
            cin>>con；
    }while(con=='Y'||con=='y');
        return 0；
}
```

程序运行结果如图 1.9.1 所示。

图 1.9.1　程序运行结果图

2. 文本文件其实就是 ASCII 文件,文件的内容是以 ASCII 的形式存放的,这样的文件有利于对字符的处理操作,便于输出。

在 C++ 语言中,要实现文件的有关操作,需要如下条件:

首先,是声明流对象;

其次,是要使用 open() 函数将文件打开,建立流与文件之间的联系;

最后,是在使用完文件时,用 close() 函数将打开的文件关闭。

程序的实现流程描述如下:

(1) 建立流对象,打开源文件和目标文件;

(2) 逐个读出源文件中的字符,将其中的数字字符写入到目的文件中;

(3) 输出源文件和目标文件中的内容;

(4) 关闭源文件和目标文件。

```cpp
#include <iostream.h>
#include <fstream.h>
#include <stdlib.h>
class CopyFile
{
public:
    CopyFile();                 //打开源文件,建立目标文件
    ~CopyFile();                //关闭源、目的文件
    void copyfiles();           //读源文件,将其中的数字字符写入到目的文件中
    void putifile();            //输出源文件内容
    void putofile();            //输出目的文件内容
private:
    fstream inf;                //用 fstream 类定义输入/输出流对象,用来操作拷贝源文件
    fstream outf;               //用 fstream 类定义输入/输出流对象,用来操作拷贝目的文件
    char file1[20];             //存放源文件路径和文件名
    char file2[20];             //存放目的文件路径和文件名
};
CopyFile::CopyFile()
{
    cout<<"源文件路径名:";
    cin>>file1;
    cout<<"目的文件路径名:";
    cin>>file2;
    //打开源文件和目的文件
    inf.open(file1,ios::in);
    if(inf.fail())
    {
```

```
        cout<<"can not open source file:"<<file1<<endl;
        exit(0);
    }
    outf.open(file2,ios::in|ios::out);
    if(outf.fail())
    {
        cout<<"can not create target file:"<<file2<<endl;
        exit(0);
    }
}
CopyFile::~CopyFile()            //关闭源文件和目的文件
{
    inf.close();
    outf.close();
}
void CopyFile::copyfiles()
{
    char ch;
    inf.seekg(0);
    //从源文件中读出字符,并将数字字符写入到目的文件中
    inf.get(ch);
    while(!inf.eof())
    {
        if(ch>='0'&&ch<='9')
            outf.put(ch);
        inf.get(ch);
    }
}
void CopyFile::putifile()
{
    char ch;
    inf.close();
    inf.open(file1,ios::in);
    inf.get(ch);
    while(!inf.eof())
    {
        cout<<ch;
        inf.get(ch);
    }
```

```
        cout<<endl;
    }
    void CopyFile::putofile()
    {
        char ch;
        outf.seekg(0);                    //使文件指针定位在文件的首位
        outf.get(ch);
        while(!outf.eof())
        {
            cout<<ch;
            outf.get(ch);
        }
        cout<<endl;
    }
    int main()
    {
        CopyFile fcopy;
        fcopy.copyfiles();
        cout<<"源文件:"<<endl;              //输出源文件内容
        fcopy.putifile();
        cout<<"目的文件:"<<endl;            //输出目的文件内容
        fcopy.putofile();
        return 0;
    }
```

程序运行结果如图1.9.2所示。

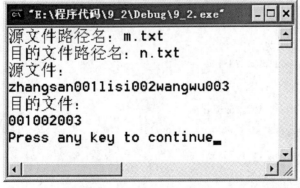

图1.9.2 程序运行结果图

3. 由于需要对学生的成绩信息进行修改，对于同一磁盘文件在程序中就可能需要频繁地进行输入和输出，因此磁盘文件的工作方式就需要被指定为输入/输出文件，即:ios::in|ios::binary。对于二进制数据文件，在每次访问时需要正确地计算好指针的定位，同时文件

中的数据要正确地进行更新。程序的实现流程可描述如下：

（1）将学生成绩信息存入文件并输出显示；

（2）输入学生学号，在磁盘文件中查询该条信息并输出，当输入为0时结束查询；

（3）输入要修改的学生信息的修改后的详细内容，将修改后的信息存入到文件中对应位置，当输入的学号为0时结束信息的修改；

（4）将修改后的文件中的所有学生的成绩信息输出显示。

```cpp
#include <iostream>
#include <fstream>
#include <iomanip>
using namespace std;
struct student                                    //定义学生成绩信息结构体
{
    int num;                                      //学号
    char name[20];                                //姓名
    int grade;                                    //班级
    float score;                                  //成绩
};
int main()
{
    student stud[10]={
    {1001,"lihua",101,86.5},
    {1002,"wanglin",103,95.5},
    {1003,"liyan",102,87.0},
    {1004,"linshi",105,98.3},
    {1005,"shilin",105,89.6},
    {1006,"weixin",104,78},
    {1007,"lili",103,78.9},
    {1008,"xiaofei",102,65.5},
    {1009,"shenyue",101,99},
    {1010,"murong",101,75.5}
    };
    int i,n;
    student stu,temp;
    fstream ofile("stud.dat",ios::out|ios::binary);    //打开一个二进制数据文件
    if(! ofile)
```

```
        {
            cerr<<"open error!"<<endl;
            exit(1);
        }
        for(i=0;i<10;i++)                           //将结构体数组中的信息存入该文件
        {
            ofile.write((char*)&stud[i],sizeof(stud[i]));
        }
        ofile.close();
        fstream ifile("stud.dat",ios::in|ios::out|ios::binary);
        cout<<setiosflags(ios::fixed)<<setprecision(2);  //定义输出浮点数的精度
        for(i=0;i<10;i++)                           //将文件中的数据读出并输出显示
        {
            ifile.read((char*)&stud[i],sizeof(stud[0]));
            cout<<stud[i].num<<setw(10)<<stud[i].name<<setw(5)<<stud[i].grade<<setw(7)<<stud[i].score<<endl;
        }
        i=0;                                        //查询指定学号的学生的成绩信息
        do
        {
            cout<<"输入要查看的学生的学号:"<<endl;
            cin>>n;
            if(n==0)break;
            ifile.seekg(i*sizeof(stud[i]),ios::beg); //查询指定学号的学生的信息位置
            ifile.read((char*)&stu,sizeof(stud[0]));
            while(n!=stu.num)
            {
                i++;
                ifile.seekg(i*sizeof(stud[i]),ios::beg);
                ifile.read((char*)&stu,sizeof(stud[0]));
            }
            cout<<stu.num<<setw(10)<<stu.name<<setw(5)<<stu.grade<<setw(7)<<stu.score<<endl;
        }while(n!=0);
        i=0;
```

```
do
{
    cout<<"输入要修改的信息:"<<endl;
    cin>>temp.num;
    if(temp.num==0)break;
    cin>>temp.name>>temp.grade>>temp.score;
    ifile.seekg(i*sizeof(stud[i]),ios::beg);        //查询指定学号的学生的信息位置
    ifile.read((char*)&stu,sizeof(stud[0]));
    while(temp.num!=stu.num)
    {
        i++;
        ifile.seekg(i*sizeof(stud[i]),ios::beg);
        ifile.read((char*)&stu,sizeof(stu));        //将修改后的信息写入文件中的相应的位置
    }
    ifile.seekg(i*sizeof(stu),ios::beg);
    ifile.write((char*)&temp,sizeof(stu));
}while(temp.num!=0);
ifile.seekg(0,ios::beg);                            //输出修改后的所有学生的成绩信息
for(i=0;i<10;i++)
{
    ifile.seekg(i*sizeof(stu),ios::beg);
    ifile.read((char*)&stu,sizeof(stu));
    cout<<stu.num<<setw(10)<<stu.name<<setw(5)<<stu.grade<<setw(7)<<stu.score<<endl;
}
ifile.close();
return 0;
}
```

程序运行结果如图 1.9.3 所示。

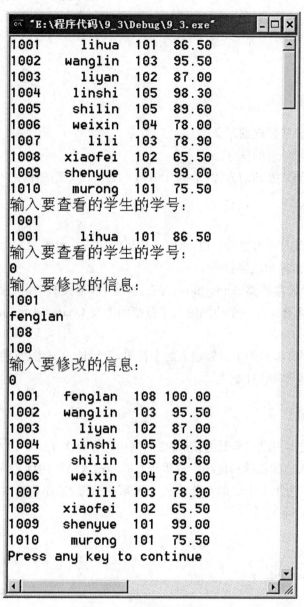

图 1.9.3 程序运行结果图

实验十　异常处理

一、实验目的

1. 理解C++中异常处理的意义。
2. 理解C++异常处理的任务。
3. 掌握C++中异常处理的方法，掌握"throw—try—catch"的使用。

二、实验内容

1. 编写程序，输入三角形的三边，求解三角形的面积。设置异常处理，对不符合构成三角形条件的输出给出提示或警告。

2. 试设计一个异常基类 Abnormity，并在这个异常基类的基础上派生出一个响应内存不足的类 Memory Error 和一个检测输入字符范围的类 CharacterError，编写程序实现并测试这三个类。

3. 建立一个字符串类用来存储输入的字符串，设计四个异常：字符串长度超出范围、字符下标越界、字符越界和程序结束。

三、实验分析

1. 我们首先提出构成三角形的条件：任意两边之和要大于第三边，三边均不为0。在这里，我们设定当输入边的长的长度小于或等于0时为输入结束的标志，此时抛出整型数异常 catch(int)。对于三边都大于0但不能构成三角形的情况，给出警告并结束程序，此时抛出实数异常 catch(double)。

```
#include<iostream>
#include<cmath>
using namespace std;
double triangle(double a,double b,double c)        //三角形面积计算函数
{
    double s=(a+b+c)/2;
    if(a+b<=c||b+c<=a||c+a<=b)throw 1.0;           //抛出异常
    return sqrt(s*(s-a)*(s-b)*(s-c));
}
int main()
{
    double a,b,c;
    try                                             //检查异常
```

```
    {
        cout<<"请输入三角形的三个边长:"<<endl;
        cout<<"a=";
        cin>>a;
        if(a<=0)throw 1;                          //抛出异常
        cout<<"b=";
        cin>>b;
        if(b<=0)throw 1;
        cout<<"c=";
        cin>>c;
        if(c<=0)throw 1;
        while(a>0&&b>0&&c>0)
        {
            cout<<"三角形的面积="<<triangle(a,b,c)<<endl;
            cout<<"请输入三角形的三个边长:"<<endl;
            cout<<"a=";
            cin>>a;
            if(a<=0)throw 1;                      //抛出异常
            cout<<"b=";
            cin>>b;
            if(b<=0)throw 1;
            cout<<"c=";
            cin>>c;
            if(c<=0)throw 1;
        }
    }
    catch(double)                                 //捕捉异常
    {
        cout<<"a="<<a<<"b="<<b<<"c="<<c<<"这不是一个三角形,异常结束!"<<endl;
    }
    catch(int)                                    //捕捉异常
    {
        cout<<"输入结束!"<<endl;
    }
    return 0;
}
```

程序运行结果如图 1.10.1 所示。

图 1.10.1　程序运行结果图

2. Abnormity 作为一个基类，可以不响应任何类型的错误，只具有一个类的通用形式，包括构造函数、析构函数和错误警告的输出。在类 Memory Error 中，对错误警告的输出函数进行重新定义，用来显示内存不足的错误警告信息。同理，对于类 CharacterError 也有相类似的处理。

```
#include<iostream>
using namespace std;
class Abnormity                          //声明异常基类
{
public:
    Abnormity(){}
    ~Abnormity(){}
    virtual void PrintError()=0;         //输出错误
};
class MemoryError:public Abnormity       //声明内存错误类
{
public:
    MemoryError(){}
    ~MemoryError(){}
    virtual void PrintError();
};
class CharacterError:public Abnormity
{
public:
    CharacterError(char ch){Badch=ch;}
    ~CharacterError(){}
```

```cpp
    virtual void PrintError();
    virtual char GetChar(){return Badch;}
    virtual void SetChar(char ch){Badch=ch;}
private:
    char Badch;
};
void MemoryError::PrintError()
{
    cout<<"内存分配时发生错误!"<<endl;
}
void CharacterError::PrintError()
{
    cout<<"输入字符非法! 你输入的是:"<<GetChar()<<endl;
}
char * memory_char()                    //为输入字符建立存储空间
{
    char * n= new char;
    if(n==0)                             //建立不成功时抛出错误
        throw MemoryError();
    return n;
}
void input_char(char * p)                //输入单个字符并判断其范围
{
    char ch;
    cout<<"请输入一个数字字符或字母:";
    cin>>ch;                             //当输入字符不在范围内时抛出错误
    if(!(((ch>='a')&&(ch<='z')||(ch>='A')&&(ch<='Z')||(ch>='0'&&ch<='9'))))
        throw CharacterError(ch);
    * p=ch;
}
void new_char()
{
    while(1)
    {
        char * p=memory_char();  input_char(p);
        cout<<"输入的字符是:"<< * p<<endl;
        delete p;
    }
}
```

```
int main()
{
    try                                    //检查错误
    {
        new_char();
    }
    catch(Abnormity &abnormal)             //捕捉错误
    {
        abnormal.PrintError();
    }
    return 0;
}
```

程序运行结果如图 1.10.2 所示。

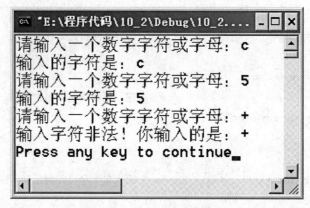

图 1.10.2　程序运行结果图

3. 对于字符串对象中存储的字符串的长度,设定一个上限值,当输入字符串的长度超出这个上限时,发生错误,利用 Extent 类来处理这一现象,在屏幕上显示一个错误。Range 类用来处理字符下标越界的异常情况,当下标值在 0 到字符串长度减 1 之外时,发生错误,Range 可返回该下标,并在屏幕上显示一个错误。Over_q 类用来处理字符越界的异常情况,当[ ]运算符重载在 String 对象中检测到一个字符——按字典顺序在'q'之后的小写字母,该异常处理程序就在屏幕上显示一个错误。End_input 类用来结束程序,当输入字符串为"quit"时,结束整个程序的运行,并输出一条信息到屏幕上。

```
# include <iostream>
# include <string>
using namespace std;
class String
{
public:
    String(char * ,int);
```

```cpp
        int Size(){return sz;}          //返回字符串长度
        class Range                     //异常类1
{
public:
        Range(int j):index(j){}
        int re_index(){return index;}
private:
        int index;
};
class Extent{};                         //异常类2
class Over_q{};                         //异常类3
char &operator[](int k)                 //重载方括号运算符
{
    if(k>=0&&k<=sz)
    {
    if(p_str[k]=='q'||(k!=sz&&p_str[k]<='z'&&p_str[k]>='q'))
        throw Over_q();
    return p_str[k];
    }
    throw Range(k);
}
private:
    char *p_str;
    int sz;
    static int max;
};
class End_input{};                      //正常结束类
int String::max=10;
String::String(char *str,int size)
{
    if(size<0||max<size)                //检查字符串长度是否超过限制
        throw Extent();
    p_str=new char[size];
    strncpy(p_str,str,size);
    sz=size;
}
void g(String &str)                     //输出字符串
{
    int num=str.Size();
```

```cpp
        for(int n=0;n<num;n++)
            cout<<str[n];
        cout<<endl;
}
void f(char *str,int len)
{
    int i;
    try                                    //检查异常
    {
        if(strcmp(str,"quit")==0)
            throw End_input();
        String s(str,len);
        cerr<<"输入第 i 个字符:";
        cin>>i;
        cout<<s[i-1]<<endl;
        g(s);
    }
    catch(String::Range r)
    {
        cerr<<endl;
        cerr<<"下标越界:"<<r.re_index()<<endl;
    }
    catch(String::Extent)
    {
        cerr<<endl;
        cerr<<"字符串长度非法!"<<endl;
    }
    catch(String::Over_q)
    {
        cerr<<endl;
        cerr<<"字符越过q."<<endl;
    }
    catch(End_input)                       //遇到 quit 结束程序
    {
        cout<<"程序结束!"<<endl;
        exit(0);
    }
    cout<<endl;
    cout<<"程序将在此处继续被执行.";
```

}
```
int main()
{
    char * ch,a[100];
    ch=a;
    while(1)                          //重复输入字符串
    {
        cout<<"输入字符串:";
        cin>>ch;
        f(ch,strlen(ch));
        cout<<endl;
    }
    return 0;
}
```

程序运行结果如图 1.10.3 示。

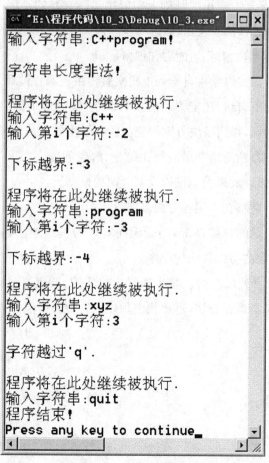

图 1.10.3　程序运行结果图

第二部分　C++课程设计

课程设计是在学习面向对象基本理论之后的实践教学环节。学生所学的知识和经验都要在这个阶段通过一个具有实际应用价值的项目开发进行检验，而这种具有更自主性的学习方式也为学生提供了一个发挥创新能力和探索更深入的学习目标的机会。课程设计的时间只有短短数周，面对一个综合性较强的设计任务，学习和工作的效率极为重要，而高效率的实现通常取决于明确的目标、合理的计划和实施过程中科学的工作方法。

C++课程设计是"C++面向对象程序设计"课程体系的组成之一，其内容是课程主干知识教学的继续，也是在学习理论知识的基础上向应用实践领域的延伸。本课程设计的目的使学生通过参加小型应用软件的开发过程，进一步掌握面向对象的程序设计方法，提高在实践中解决问题的能力，培养学生的创新能力和创新意识。

本课程设计的教学形式以学生自学和上机实验为主。课程设计过程中采用研究型学习方法。学生在教师指导下，自行选定力所能及的课题，学习运用科学研究的理念、模式和方法分析所选课题中的问题，提出解决方案。充分发挥学生的主动性和创造性，通过检索资料对资料经行分析、讨论、概括总结并最终得出结论、解决问题。要求学生独立完成相应软件设计文档的撰写，完成设计成果的测试以及撰写课程设计报告。

学生通过课程设计能够在下述各方面得到锻炼：

（1）能根据实际问题的具体情况，结合面向对象的基本理论和基本技巧，正确分析问题，并能设计出解决问题的有效算法与程序。

（2）提高程序设计和调试能力。学生通过上机实习，验证自己设计的算法和程序的正确性。学会有效利用基本调试方法，迅速找出程序代码中的错误并且修改，进一步提高程序设计水平。

样例一 学生成绩管理系统

一、设计的任务要求

设计一个简易的学生成绩管理系统,能够完成学生成绩的增加、删除、查找、修改、统计等操作,数据信息使用文件保存。要求系统具有菜单和提示,界面友好。

二、程序功能设计

1. 设计程序功能

学生成绩系统中学生的成绩信息按照学号的顺序进行存放。根据任务要求,下面将系统功能进行详细划分,功能结构如图2.1.1所示。

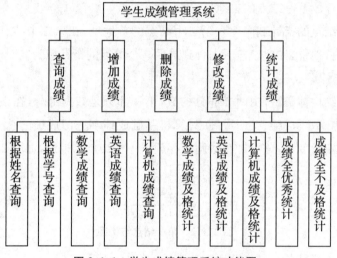

图 2.1.1 学生成绩管理系统功能图

成绩的增加:通过键盘输入学生成绩信息并将其添加到学生成绩信息记录中,要求按照学号顺序插入。

成绩的删除:根据学生的学号从学生成绩信息记录中删除该学生成绩信息。

成绩的查询:可以根据学生的学号和姓名查询学生成绩信息,也可以根据某一门成绩的分数段查询学生成绩信息。

成绩的修改:根据学生的学号修改相应学生的成绩信息。

成绩的输出:将所有学生成绩信息输出。

成绩的统计:统计每门课程的及格人数显示不及格学生的信息,统计三门课程成绩全为优秀的学生人数,显示三门课程成绩全不及格的学生信息。

保存数据：利用文件操作将链表中学生成绩信息保存到文件。

加载数据：利用文件操作从文件中读取学生成绩信息，形成学生记录的链表。

2. 设计数据格式

要完成学生成绩信息的增删改查及统计，首先设计一下内存中存放数据信息的格式。在本设计中采用动态内存空间分配的链表方法，该方法为一个结构分配内存空间。每一次分配一块空间可用来存放一个学生成绩的数据，可称之为一个结点。有多少个学生就应该申请分配多少块内存空间，也就是说要建立多少个结点。当然用结构数组也可以完成上述工作，但如果预先不能准确把握学生人数，也就无法确定数组大小。而且当学生留级、退学之后也不能把该元素占用的空间从数组中释放出来。

用动态存储的方法可以很好地解决这些问题。有一个学生就分配一个结点，无须预先确定学生的准确人数，某学生退学，可删去该结点，并释放该结点占用的存储空间，从而节约了内存资源。另一方面，用数组的方法必须占用一块连续的内存区域。而使用动态分配时，每个结点之间可以是不连续的(结点内是连续的)。结点之间的联系可以在结点结构中定义一个指针项用来存放下一结点的首地址。

可在第一个结点的指针域内存入第二个结点的首地址，在第二个结点的指针域内又存放第三个结点的首地址，如此串连下去直到最后一个结点的指针域为空。

3. 设计结点的组成

结点是一个结构体类型，由两个部分构成，第一部分是数据部分，在该系统中用来存放学生的学号、姓名、数学成绩、英语成绩、计算机基础成绩和三门成绩的总分；第二个部分定义了一个结构体指针变量，该指针变量用来存放下一个结点的地址，如图2.1.2所示。

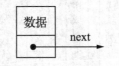

图 2.1.2　单个结点结构图

定义存放学生成绩信息结点的语句如下：

```
struct score              //定义存放学生成绩信息的结点
{
    int num;              //学号
    string name;          //姓名
    float math;           //数学成绩
    float english;        //英语成绩
    float computer;       //计算机基础成绩
    float scoresum;       //三门成绩总和
    struct score * next;  //next为指向下一结点的指针
};
```

一个个结点就通过 next 指针连接起来形成了单向链表的结构,如图 2.1.3 所示。head 指针指向链表的首结点。

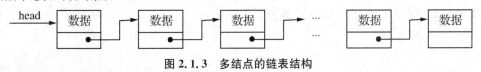

图 2.1.3 多结点的链表结构

4. 程序设计与实现

在本程序中,使用链表存放学生成绩数据,设计一个功能类 record 来完成系统的各项功能。具体设计如下:

```
class record
{
public:
    struct score * InsertRecord(struct score *h);    //增加学生成绩信息
    struct score * DeleteRecord(struct score *h);    //删除学生成绩信息
    struct score * UpdateRecord(struct score *h);    //修改学生成绩信息
    void FindRecord(struct score *h,int x,float s1,float s2);
    //根据某门课程的分数段查询学生成绩信息
    void FindRecord(struct score *h,string  x);
    //根据学生姓名查询学生成绩信息
    void FindRecord(struct score *h,int  x);         //根据学生学号查询学生成绩信息
    void StatisticRecord(struct score *h,int x);
    //统计某门课程的及格学生人数、及格率,并显示不及格学生信息
    void StacRecordFine(struct score *h);
    //统计三门课程成绩全为优秀的学生人数,并显示全为优秀的学生信息
    void StacRecordDisq(struct score *h);
    //统计三门课程成绩全部不及格的学生人数,并显示全部不及格的学生信息
    void PrintRecord(struct score *h);               //输出所有学生成绩信息
    void SaveRecordFile(struct score *h);            //保存学生成绩信息到文件
    struct score * LoadRecordFile(struct score *h);
    //从文件中加载学生成绩信息
};
```

(1) 链表的插入操作

链表的插入操作是针对有序链表说的,本程序是按照学生的学号从大到小顺序存放链表中结点的。新结点要插入链表前,应首先找到结点要插入的位置,本程序中就是查找第一个比要插入新结点学号大的结点,将新结点插入到找到的第一个学号比其大的结点之前。在本功能模块中约定指针 h 为链表的头指针,指针 p1 为查找的符合要求的结点,指针 p2 为 p1 的前一个结点,指针 p3 为新增加的结点。根据链表是否为空及找到结点的位置情况,链表的插入操作需要考虑下面四种情况。

● 链表为空

首先考虑链表中没有结点的情况,也就是链表为空,这时新增加的结点就是唯一的结点,新增结点就作为链表的头结点。具体语句如下:

```
if(h==NULL)
    {
        h=p3;              //将新结点 p3 作为链表的头结点
        return h;
    }
```

● 插入到原链表头结点之前

接着考虑链表中有结点时,即链表不为空时,我们需要查找插入点的位置,这需要通过循环从头到尾扫描链表中的每个结点,当找到一个结点的学号大于新结点的学号或者查找结点为空为止。

若经过循环查找到的第一个大于新结点的学号的结点正好是链表头结点时,将新的结点 p3 插入到头结点 h 之前,再将 p3 作为新的链表头结点,如图 2.1.4 所示。

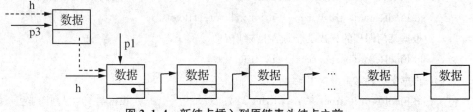

图 2.1.4 新结点插入到原链表头结点之前

具体语句如下:

```
if(p1==h)
{
    p3->next=h;            //新结点的指针域指向链表的头结点
    h=p3;                  //将新结点作为链表的头结点
    return h;
}
```

● 插入到链表中间的位置

若经过循环查找到第一个大于新结点学号的结点是除头结点以外的其他结点时,将新的结点 p3 插入到找到的结点 p1 之前,即将指针 p2 的 next 指向新结点 p3,将 p3 结点的 next 指向 p1 指针所指向的结点,如图 2.1.5 所示。

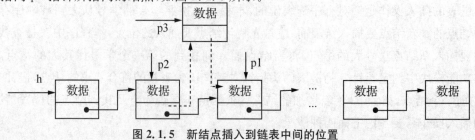

图 2.1.5 新结点插入到链表中间的位置

具体语句如下:
```
if(p3->num<=p1->num)
{
    p2->next=p3;
    p3->next=p1;
}
```

● 插入到链表尾部

若经过循环查找没有找到大于新结点学号的结点,即p1为空指针,则将新结点插入链表的最后一个结点的后面作为链表的新的末尾结点。具体操作为将p2结点的指针域设置为p3,将新的末尾结点p3的指针域设置为空即可赋值为p1,此操作和新结点插入到链表中间的位置的代码类似,所以将新结点插入到链表尾部和插入到链表中间位置的操作合并在一起,在此不再说明。

(2) 链表的删除操作

删除链表,首先根据学号在链表中通过循环查找与要删除学生的学号相同的结点,若找到则在链表中删除查找到符合条件的结点,并将该结点的空间释放,若没找到给出相应的提示。在本功能模块中约定指针h为链表的头指针,指针p1为查找的符合要求的结点的指针,指针p2为p1的前一个结点指针。根据链表是否为空及找到符合要求结点的位置,链表的删除操作需要考虑下面三种情况。

● 链表为空

首先考虑链表中没有结点的情况,也就是链表为空,就没有结点可以删除了,可以给出适当的提示。具体语句如下:
```
if(h==NULL)
{
    cout<<"\n抱歉,没有任何记录!";
    return h;
}
```

● 删除的结点为链表头结点

当查找到的需删除的结点为链表头结点时,即查找到符合条件的结点指针p1指向的就是链表的头结点h,则将头结点指针h指向下一个结点,将下一个结点作为链表的头结点,并且释放原来头结点的存储单元,如图2.1.6所示。

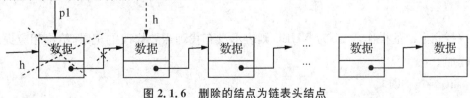

图2.1.6 删除的结点为链表头结点

具体语句如下：
```
if(p1==h)
{    h=h->next;
     delete p1;
}
```

● 删除的结点为链表中非头结点的其他结点

当查找到的需删除的结点 p1 不是链表头结点 h 时，则将 p2 指向的结点的指针域指向 p1 指向结点的后一个结点，并且释放 p1 结点的存储单元，如图 2.1.7 所示。

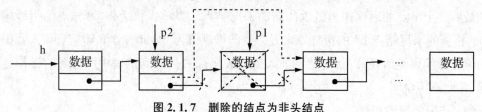

图 2.1.7　删除的结点为非头结点

具体语句如下：

p2->next=p1->next;

delete p1;

（3）链表的更新操作

要修改链表中结点的记录，则首先在链表中通过循环查找需要修改的记录结点，若找到则根据要求修改结点的相关数据，没有找到符合条件的结点则给出没有找到的提示。

在本功能模块中，约定指针 h 为链表的头指针，指针 p1 为查找的符合要求的结点。查找要修改的结点，首先将头结点指针 h 赋值给指针 p1，若 p1 不为 NULL 且 p1 的学号不等于需要修改的学生学号，则将 p1 指针移向下一个结点，若下移后 p1 不等于 NULL 且 p1 的学号不等于需要修改的学生学号，则再将 p1 指针移向下一个结点，重复执行操作，直到 p1 为空或 p1 的学号等于需要修改的学生学号为止。具体代码如下：

```
p1=h;
    cout<<"\n请输入要修改记录的学生学号";
    cin>>num;
    while(p1!=NULL&&p1->num!=num)
    {
        p1=p1->next;
    }
```

执行完上面的操作若 p1 为 NULL 则表示没有找到需要修改的记录，给出相应提示，具体代码如下：

```
if(p1==NULL)
{
```

```
        cout<<"\n 抱歉啊,表中没有该记录的哦!";
        return h;
}
```

相反则找到需要修改的记录结点,即 p1 指向的结点,根据需要进行修改,具体代码如下:

```
        cout<<"\n 请重新输入学生的数学成绩:";
        cin>>p1->math;
        cout<<"\n 请重新输入学生的英语成绩:";
        cin>>p1->english;
        cout<<"\n 请重新输入学生的计算机基础成绩:";
        cin>>p1->computer;
        p1->scoresum=p1->math+p1->english+p1->computer;
```

(4) 链表的输出操作

链表的输出操作,就是通过循环扫描链表中的每一个结点,将结点的数据域的信息输出。在本功能中约定指针 p 为指向当前结点的指针。首先将指针 p 指向链表的头结点,若 p 不为空则输出指针 p 指向的各项数据,而后将指针 p 移向下一个结点,若 p 再不为空则再输出指针 p 指向的各项数据,接着再将指针 p 移向下一结点,依次循环直到指针 p 为空为止。因链表的输出操作相对简单,具体代码可以查看源程序,不再书写。

5. 程序设计的完整源代码以及注释

```
#include<iostream>
#include<string>
#include<fstream>
using namespace std;
struct SCORE                    //定义存放学生成绩信息的结点
{
        int num;                //学号
        string name;            //姓名
        float math;             //数学成绩
        float english;          //英语成绩
        float computer;         //计算机基础成绩
        float scoresum;         //总成绩
        struct SCORE * next;    //next 为指向下一结点的指针
};
struct SCORE * head;            //指向链表头结点的指针
int studentSum=0;               //学生总人数
class record
```

```cpp
{
public:
    struct SCORE * InsertRecord(struct SCORE *h);   //增加学生成绩信息
    struct SCORE * DeleteRecord(struct SCORE *h);   //删除学生成绩信息
    struct SCORE * UpdateRecord(struct SCORE *h);   //修改学生成绩信息
    void FindRecord(struct SCORE *h,int x,float s1,float s2);
    //根据某门课程的分数段查询学生成绩信息
    void FindRecord(struct SCORE *h,string x);//根据学生姓名查询学生成绩信息
    void FindRecord(struct SCORE *h,int x);//根据学生学号查询学生成绩信息
    void StatisticRecord(struct SCORE *h,int x);
    //统计某门课程的及格学生人数、及格率,并显示不及格学生信息
    void StacRecordFine(struct SCORE *h);
    //统计三门课程成绩全为优秀的学生人数,并显示全为优秀的学生信息
    void StacRecordDisq(struct SCORE *h);
    //统计三门课程成绩全部不及格的学生人数,并显示全部不及格的学生信息
    void PrintRecord(struct SCORE *h);              //输出所有学生成绩信息
    void SaveRecordFile(struct SCORE *h);           //保存学生成绩信息到文件
    struct SCORE * LoadRecordFile(struct SCORE *h);//从文件中加载学生成绩信息
};
struct SCORE * record::InsertRecord(struct SCORE *h)
{
    struct SCORE *p1,*p2,*p3;          //定义指针变量 p1、p2、p3
    p3=new SCORE;                      //创建新的学生成绩结点
    cout<<"\n 请输入学生学号:";
    cin>>p3->num;                      //从键盘接受输入数据赋值给结点的学号
    cout<<"\n 请输入学生姓名:";
    cin>>p3->name;                     //从键盘接受输入数据赋值给结点的姓名
    cout<<"\n 请输入学生的数学成绩:";
    cin>>p3->math;                     //从键盘接受输入数据赋值给结点的数学成绩
    cout<<"\n 请输入学生的英语成绩:";
    cin>>p3->english;                  //从键盘接受输入数据赋值给结点的英语成绩
    cout<<"\n 请输入学生的计算机基础成绩:";
    cin>>p3->computer;                 //从键盘接受输入数据赋值给结点的计算机成绩
    p3->scoresum=p3->math+p3->english+p3->computer;//计算结点的总成绩
    p3->next=NULL;                     //将要插入结点的指针域设置为空
    if(h==NULL)                        //当链表中没有结点时,将要新插入的结点作为头结点
    {
```

```
        h=p3;
        return h;
}
p1=p2=h;
while(p1!=NULL&&p3->num>p1->num)
//查找结点的学号大于要插入结点学号的第一个结点
//指针 p1 表示符合条件的结点的指针,指针 p2 是指针 p1 的前一个结点指针
{
        p2=p1;
        p1=p1->next;
}
if(p1==h)                          //插入位置为头结点前
{
        p3->next=h;
        h=p3;
        return h;
}
else                               //插入位置为链表的中间和链表尾部
{
        p2->next=p3;
        p3->next=p1;
}
studentSum+=1;                     //学生人数加 1
return h;//返回链表的头结点
}


void record::PrintRecord(SCORE *h)
{
    if(h==NULL)
    {
        cout<<"\n抱歉,没有任何记录!\n";
        return;
    }
    cout<<"\n学号\t姓名\t数学\t英语\t计算机\t总分"<<endl;
    while(h)                       //输出链表中每个结点的学生成绩信息
    {
        cout<<h->num<<"\t"<<h->name<<"\t"<<h->math<<"\t"<<h->english<
```

```cpp
<"\t"<<h->computer<<"\t"<<h->scoresum<<endl;
            h=h->next;
        }
    }
    struct SCORE * record::DeleteRecord(struct SCORE * h)
    {
        struct SCORE *p1,*p2;
        int num;
        if(h==NULL)                    //链表为空
        {
            cout<<"\n抱歉,没有任何记录!";
            return h;
        }
        p1=p2=h;                       //将链表的头结点指针h赋值给指针p1和指针p2
        cout<<"\n请输入要删除记录的学生学号";
        cin>>num;
        while(p1!=NULL&&p1->num!=num)
        //查找结点的学号等于要删除学生学号的第一个结点
        //指针p1表示符合条件的结点的指针,指针p2是指针p1的前一个结点指针
        {
            p2=p1;
            p1=p1->next;
        }
        if(p1==NULL)                   //没有找到符合要求的结点
        {
            cout<<"\n抱歉啊,表中没有该记录的哦!";
            return h;
        }
        if(p1->num==num)
        {
            studentSum-=1;             //学生人数减1
            if(p1==h)                  //删除的是头结点
               h=h->next;
            else                       //删除的是非头结点
               p2->next=p1->next;
            delete p1;                 //释放p1所指向的存储单元
        }
```

```cpp
        return h;
}
struct SCORE * record::UpdateRecord(struct SCORE * h)
{
        struct SCORE * p1;
        int num;
        if(h==NULL)//链表为空
        {
                cout<<"\n 抱歉,没有任何记录!";
                return h;
        }
p1=h;                         //将链表的头结点指针 h 赋值给指针 p1
cout<<"\n 请输入要修改记录的学生学号";
cin>>num;
while(p1!=NULL&&p1->num!=num)
//查找结点的学号等于要修改学生学号的结点指针
{
    p1=p1->next;//将 p1 指针移到下一个结点
}
        if(p1==NULL)//没有找到符合要求的结点
{
        cout<<"\n 抱歉啊,表中没有该记录的哦!";
        return h;
}
if(p1->num==num)              //找到符合要求的结点,并修改学生的相关成绩
{
        cout<<"\n 请重新输入学生的数学成绩:";
        cin>>p1->math;
        cout<<"\n 请重新输入学生的英语成绩:";
        cin>>p1->english;
        cout<<"\n 请重新输入学生的计算机基础成绩:";
        cin>>p1->computer;
        p1->scoresum=p1->math+p1->english+p1->computer;
}
        return h;
}
```

```cpp
void record::FindRecord(struct SCORE *h,int x,float s1,float s2)
{
    if(h==NULL)                              //链表为空
    {
        cout<<"\n抱歉,没有任何记录!";
        return ;
    }
    cout<<"\n学号\t姓名\t数学\t英语\t计算机\t总分"<<endl;
    while(h)
    {
        if(x==1)                             //查找数学成绩在某分数段的学生成绩信息
            if(h->math>=s1&&h->math<=s2)
                cout<<h->num<<"\t"<<h->name<<"\t"<<h->math<<"\t"<<h->english<<"\t"<<h->computer<<"\t"<<h->scoresum<<endl;
        if(x==2)                             //查找英语成绩在某分数段的学生成绩信息
            if(h->english>=s1&&h->english<=s2)
                cout<<h->num<<"\t"<<h->name<<"\t"<<h->math<<"\t"<<h->english<<"\t"<<h->computer<<"\t"<<h->scoresum<<endl;
        if(x==3)                             //查找计算机成绩在某分数段的学生成绩信息
            if(h->computer>=s1&&h->computer<=s2)
                cout<<h->num<<"\t"<<h->name<<"\t"<<h->math<<"\t"<<h->english<<"\t"<<h->computer<<"\t"<<h->scoresum<<endl;
        h=h->next;
    }
}
void record::FindRecord(struct SCORE *h,int num)    //根据学生学号查找学生成绩信息
{
    struct SCORE *p1;
    if(h==NULL)                              //链表为空
    {
        cout<<"\n抱歉,没有任何记录!";
        return ;
    }
    p1=h;                                    //将链表的头结点指针h赋值给指针p1
    while(p1!=NULL&&p1->num!=num)
    //查找结点的学号等于要查找学生学号的结点指针
    {
```

```cpp
        p1=p1->next;
    }
    if(p1==NULL)                        //没有找到
    {
        cout<<"\n 抱歉啊,表中没有该记录的哦!";
        return;
    }
    if(p1->num==num)                    //找到并显示信息
    {
        cout<<"\n 学号\t 姓名\t 数学\t 英语\t 计算机\t 总分"<<endl;
        cout<<p1->num<<"\t"<<p1->name<<"\t"<<p1->math<<"\t"<<p1->english<<"\t"<<p1->computer<<"\t"<<p1->scoresum<<endl;
    }
}
void record::FindRecord(struct SCORE *h,string name)  //根据学生姓名查找学生成绩信息
{
    struct SCORE *p1;
    if(h==NULL)                         //链表为空
    {
        cout<<"\n 抱歉,没有任何记录!";
        return;
    }
    p1=h;                               //将链表的头结点指针 h 赋值给指针 p1
    while(p1!=NULL&&p1->name!=name)
    //查找结点的姓名等于要查找学生姓名的结点指针
    {
        p1=p1->next;
    }
    if(p1==NULL)                        //没有找到符合要求的结点
    {
        cout<<"\n 抱歉啊,表中没有该记录的哦!";
        return;
    }
    if(p1->name==name)                  //找到符合条件的结点并显示信息
    {
        cout<<"\n 学号\t 姓名\t 数学\t 英语\t 计算机\t 总分"<<endl;
        cout<<p1->num<<"\t"<<p1->name<<"\t"<<p1->math<<"\t"<<p1->
```

```cpp
english<<"\t"<<p1->computer<<"\t"<<p1->scoresum<<endl;
        }
    }
    void record::StatisticRecord(struct SCORE *h,int x)
    {
        struct SCORE *p=h;//将链表的头结点指针 h 赋值给指针 p
        int count=0;//定义统计人数 count 变量并赋初值为 0
        if(p==NULL)//链表为空
        {
            cout<<"\n抱歉,没有任何记录!";
            return ;
        }
        while(p)
        {
            if(x==1)                        //统计数学成绩及格的学生人数
                if(p->math>=60)
                    count+=1;
            if(x==2)                        //统计英语成绩及格的学生人数
                if(p->english>=60)
                    count+=1;
            if(x==3)                        //统计计算机成绩及格的学生人数
                if(p->computer>=60)
                    count+=1;
            p=p->next;
        }
        if(x==1)                            //显示数学成绩及格的学生人数及及格率
        {
            cout<<"数学成绩及格学生人数为";
            cout<<count;
            cout<<",及格率为";
            cout<<count/(float)studentSum<<endl;
            if(count<studentSum)
                cout<<"\n学号\t姓名\t数学"<<endl;
            else
                cout<<"没有数学成绩不及格学生"<<endl;
        }
        else
```

```cpp
        if(x==2)                    //显示英语成绩及格的学生人数及及格率
        {
        cout<<"英语成绩及格学生人数为";
        cout<<count;
        cout<<",及格率为";
        cout<<count/(float)studentSum<<endl;
        if(count<studentSum)
           cout<<"\n学号\t姓名\t英语"<<endl;
        else
           cout<<"没有英语成绩不及格学生"<<endl;
        }
        else
          if(x==3)                  //显示计算机成绩及格的学生人数及及格率
          {
              cout<<"计算机成绩及格学生人数为";
              cout<<count;
              cout<<",及格率为";
              cout<<count/(float)studentSum<<endl;
              if(count<studentSum)
                  cout<<"\n学号\t姓名\t计算机"<<endl;
              else
                  cout<<"没有计算机成绩不及格学生"<<endl;
          }
        p=h;
        while(p)
        {
              if(x==1)              //显示数学成绩不及格的学生信息
                  if(p->math<60)
                      cout<<p->num<<"\t"<<p->name<<"\t"<<p->math<<endl;
              if(x==2)              //显示英语成绩不及格的学生信息
                  if(p->english<60)
                      cout<<p->num<<"\t"<<p->name<<"\t"<<p->english<<endl;
              if(x==3)              //显示计算机成绩不及格的学生信息
                  if(p->computer<60)
                      cout<<p->num<<"\t"<<p->name<<"\t"<<p->computer<
```

```cpp
        <endl;
                p=p->next;
            }
        }
        void record::StacRecordFine(struct SCORE *h)
        {
            struct SCORE *p=h;          //将链表的头结点指针 h 赋值给指针 p
            int count=0;                //定义统计人数 count 变量并赋初值为 0
            if(p==NULL)                 //链表为空
            {
                cout<<"\n抱歉,没有任何记录!";
                return ;
            }
            while(p)
            {
                if(p->math>=90&&p->english>=90&&p->computer>=90)
                    //统计三门成绩全部为优秀的学生人数
                    count+=1;
                p=p->next;              //将指针移到下一结点
            }
            cout<<"三门成绩全为优秀的学生人数为";
            cout<<count<<endl;
            cout<<"全为优秀的学生信息为:"<<endl;
            cout<<"\n学号\t姓名\t数学\t英语\t计算机\t总分"<<endl;
            p=h;                        //将链表的头结点指针 h 赋值给指针 p
            while(p)
            {
                if(p->math>=90&&p->english>=90&&p->computer>=90)
                    //显示三门成绩全部为优秀的学生信息
        cout<<p->num<<"\t"<<p->name<<"\t"<<p->math<<"\t"<<p->english<<"\t"<<p->computer<<"\t"<<p->scoresum<<endl;
                p=p->next;
            }
        }
        void record::StacRecordDisq(struct SCORE *h)
        {
            struct SCORE *p=h;//将链表的头结点指针 h 赋值给指针 p
```

```cpp
    int count=0;//定义统计人数的count变量并赋初值为0
    if(p==NULL)//链表为空
    {
            cout<<"\n抱歉,没有任何记录!";
            return ;
    }
    while(p)
    {
            if(p->math<60&&p->english<60&&p->computer<60)
                //统计三门成绩全部不及格的学生人数
                count+=1;
            p=p->next;
    }
    cout<<"三门成绩全部不及格的学生人数为";
    cout<<count<<endl;
    cout<<"全为不及格的学生信息为:"<<endl;
    cout<<"\n学号\t姓名\t数学\t英语\t计算机\t总分"<<endl;
    p=h;
    while(p)
    {
            if(p->math<60&&p->english<60&&p->computer<60)
                //显示三门成绩全部不及格的学生信息
cout<<p->num<<"\t"<<p->name<<"\t"<<p->math<<"\t"<<p->english<<"\t"<<p->computer<<"\t"<<p->scoresum<<endl;
            p=p->next;
    }
}
void record::SaveRecordFile(struct SCORE *h)//将链表中的数据写入文件
{
struct SCORE *p;
ofstream ofile;                  //定义输出文件对象
ofile.open("score.dat",ios::out);
//以写的方式打开文件score.dat,若该文件不存在,则创建score.dat文件
if(!ofile)                       //文件打开错误
{
        cout<<"\n数据文件打开错误没有将数据写入文件!\n";
        return ;
```

```cpp
        }
        ofile<<"学号\t姓名\t数学\t英语\t计算机\t总分";
        while(h)
        {
            ofile<<endl<<h->num<<"\t"<<h->name<<"\t"<<h->math<<"\t"<<h->english<<"\t"<<h->computer<<"\t"<<h->scoresum;
            //将当前结点的数据信息写入到文件中
            p=h;h=h->next;
            delete p;
        }
        ofile.close();                                  //关闭文件对象
}
struct SCORE * record::LoadRecordFile(struct SCORE * h)
{
        ifstream ifile;                                 //定义输入文件对象
        ifile.open("score.dat",ios::in);                //以读的方式打开文件 score.dat
        struct SCORE * p,* q;
        if(! ifile)                                     //文件打开错误
        {
            cout<<"\n 数据文件不存在,加载不成功! \n";
            return NULL;
        }
        char s[50];
        ifile.getline(s,50);                            //读取文件指针当前行数据
        while(! ifile.eof())
        {
            studentSum=studentSum+1;                    //学生人数加 1
            p=new SCORE;                                //创建新的 score 变量
            ifile>>p->num>>p->name>>p->math>>p->english>>p->computer>>p->scoresum;
            //将数据从文件中读取到新的结点中
            p->next=NULL;                               //新结点的指针域为空
            if(h==NULL)                                 //将新结点插入到链表中
                q=h=p;
            else
            {
                q->next=p;
```

```cpp
            q=p;
        }
    }
    ifile.close();                              //关闭文件对象
    return h;
}
void SystemMenu(record r)                       //系统菜单，及处理用户的选择
{
    int choice;
    while(1)
    {
        cout<<"\n\t\t 欢迎进入学生成绩管理系统！";   //显示系统主菜单
cout<<"\n@@@@@@@@@@@@@@@@@@@@@@@@@@@@@@@@@@@@@@@@@@@@@@@@@@";
        cout<<"\n\t1、添加学生成绩信息";
        cout<<"\n\t2、删除学生成绩信息";
        cout<<"\n\t3、修改学生成绩信息";
        cout<<"\n\t4、查询学生成绩信息";
        cout<<"\n\t5、显示所有学生成绩信息";
        cout<<"\n\t6、统计学生成绩信息";
        cout<<"\n\t0、退出系统";
cout<<"\n@@@@@@@@@@@@@@@@@@@@@@@@@@@@@@@@@@@@@@@@@@@@@@@@@@";
        cout<<"\n 请根据提示选择操作：";
        cin>>choice;
        switch(choice)
        {
        case 1:                                 //增加学生成绩信息
            head=r.InsertRecord(head);
            break;
        case 2:                                 //删除学生成绩信息
            head=r.DeleteRecord(head);
        case 3:                                 //修改学生成绩信息
            head=r.UpdateRecord(head);
        case 4:                                 //查询学生成绩信息
            while(1)
            {
```

```cpp
        int c;
        cout<<"\n* * * * * * * * * * * * * * * * * * * * * * * * * * * * * * * * * * * * * * * * * * * * * *";
        cout<<"\n\t1、根据学号查询学生成绩信息";
        cout<<"\n\t2、根据姓名查询学生成绩信息";
        cout<<"\n\t3、根据数学分数查询学生成绩信息";
        cout<<"\n\t4、根据英语成绩查询学生成绩信息";
        cout<<"\n\t5、根据计算机基础成绩查询学生成绩信息";
        cout<<"\n\t6、返回上级目录";
        cout<<"\n* * * * * * * * * * * * * * * * * * * * * * * * * * * * * * * * * * * * * * * * * * * * * *";
        //显示查询子菜单
        cout<<"\n请根据提示选择操作:";
        cin>>c;
        if(c==1)                        //根据学生学号查询学生成绩信息
        {
            int x;
            cout<<"\n请输入需要查询的学生学号:";
            cin>>x;
            r.FindRecord(head,x);
        }
        if(c==2)                        //根据学生姓名查询学生成绩信息
        {
            string name;
            cout<<"\n请输入需要查询的学生姓名:";
            cin>>name;
            r.FindRecord(head,name);
        }
        if(c==3)                        //根据数学分数段查询学生成绩信息
        {
            float s1,s2;
            cout<<"\n请输入查询的数学最低分和最高分";
            cin>>s1>>s2;
            r.FindRecord(head,1,s1,s2);
        }
        if(c==4)                        //根据英语分数段查询学生成绩信息
        {
```

```cpp
        float s1,s2;
        cout<<"\n请输入查询的英语最低分和最高分";
        cin>>s1>>s2;
        r.FindRecord(head,2,s1,s2);
    }
    if(c==5)                                    //根据计算机分数段查询学生成绩信息
    {
        float s1,s2;
        cout<<"\n请输入查询的计算机基础最低分和最高分";
        cin>>s1>>s2;
        r.FindRecord(head,3,s1,s2);
    }
    if(c==6)                                    //退出查询子菜单
        break;
    }
    break;
case 5:                                         //输出所有学生成绩信息
    r.PrintRecord(head);
    break;
case 6:                                         //统计学生成绩信息
    while(1)
    {
        int c;
        cout<<"\n* * * * * * * * * * * * * * * * * * * * * * * * * * * * * * * * * * * *";
        cout<<"\n\t1、统计数学成绩及格学生人数,并显示不及格学生信息";
        cout<<"\n\t2、统计英语成绩及格学生人数,并显示不及格学生信息";
        cout<<"\n\t3、统计计算机成绩及学生格人数,并显示不及格学生信息";
        cout<<"\n\t4、统计三门功课都不及格的学生人数,并显示学生信息";
        cout<<"\n\t5、统计三门功课都优秀的学生人数,并显示学生信息(>=90)";
        cout<<"\n\t6、返回上级目录";
        cout<<"\n* * * * * * * * * * * * * * * * * * * * * * * * * * * * * * * * * * * *";
        //显示统计子菜单
        cout<<"\n请根据提示选择操作:";
        cin>>c;
```

```
            if(c==1)                    //统计数学成绩及格学生人数,并显示不及格学生信息
            {
               r.StatisticRecord(head,1);
            }
            if(c==2)                    //统计英语成绩及格学生人数,并显示不及格学生信息
            {
               r.StatisticRecord(head,2);
            }
            if(c==3)                    //统计计算机成绩及学生格人数,并显示不及格学生信息
            {
               r.StatisticRecord(head,3);
            }
              if(c==4)                  //统计三门功课都不及格的学生人数,并显示学生信息
              {
                 r.StacRecordFine(head);
              }
              if(c==5)                  //统计三门功课都优秀的学生人数,并显示学生信息
              {
                 r.StacRecordDisq(head);
              }
              if(c==6)                  //退出统计子菜单
                 break;
           }
           break;
        }
        if(choice==0)                   //退出系统
           break;
     }
}
int main()
{
      head=NULL;
      record r;                         //定义 record 类的对象 r
      head=r.LoadRecordFile(head);      //将文件中的数据读取到链表中
      SystemMenu(r);                    //显示系统菜单,并处理用户的选择
      r.SaveRecordFile(head);           //将链表中的数据写到文件中
      return 0;
}
```

样例二 通讯录管理系统

一、设计的任务要求

设计一个通讯录管理系统,能够完成通讯录记录的增加、删除、查找、修改等操作,数据信息使用文件保存。要求系统具有菜单和提示,界面友好。

二、程序功能设计

通讯录管理系统是针对储存用户联系方式以及一些简单个人信息的实用管理系统,它可以让用户对众多朋友、同学、同事、家人等信息进行储存、修改和快速查阅。根据任务的要求,下面对系统功能进行详细划分,功能结构如图 2.2.1 所示。

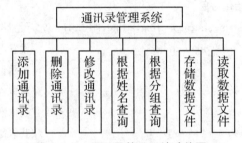

图 2.2.1 通讯录管理系统功能图

添加通讯录:按顺序将姓名(name)、性别(sex)、电子邮箱(E-mail)、地址(address)、电话(telnumber)、分组(type)依次输入生成结点,并建立链表将其连接,直到姓名输入为"0"终止。

删除通讯录:分为删除特定姓名的通讯录中的记录和删除通讯录中的全部记录。

修改通讯录:根据姓名修改通讯录中的记录。

查询通讯录:根据姓名和分组查询通讯录中的记录,并将查询结果输出。

存储数据文件:链表中的信息以文件形式被长期保存。

读取数据文件:读取文件中的数据,并将其建立链表。

三、程序设计的完整源代码以及注释

```
#include<iostream>
#include<fstream>
#include<string>
#include<iomanip>
using namespace std;
```

```cpp
class CPhoneRecord                    //定义结点的数据部分
{
private:
    int id;                           //记录编号
    string name;                      //姓名
    string sex;                       //性别
    string email;                     //邮箱
    string address;                   //住址
    string telnumber;                 //电话号码
    int type;                         //分组类型1为朋友,2为同学,3为家人,4为同事
    static int s;                     //静态成员数据s,处理记录编号
public:
    CPhoneRecord();
    CPhoneRecord(string ,string,string,string,string,int );
    //构造函数,初始化对象
    void SetRecord(string ,string,string,string,string,int);//设置电话簿记录的基本信息
    void SetTelNumber(string number);         //设置电话号码
    void Display();                           //显示信息
    string GetName();                         //获取姓名
    string GetSex();                          //获取性别
    string GetEmail();                        //获取邮箱
    string GetAddress();                      //获取住址
    string GetTelnumber();                    //获取电话号码
    int GetType();                            //获取分组类别
private:
    string TypeString(int type);    //工具函数,将类型由数字转化为字符串
};
CPhoneRecord::CPhoneRecord()
{
    s++;                            //每创建该类一个新对象,静态成员s加1
    id=s;
    //利用静态成员处理记录的编号,实现该类对象的记录编号自动递增
    name="\0";
    sex="\0";
    email="\0";
    address="\0";
    telnumber="\0";
    type=1;                         //默认分组类型设为朋友
```

}
CPhoneRecord::CPhoneRecord(string name1,string sex1,string email1,string address1,string tn, int type1)
{
    s++;　　　　　　//每创建该类一个新对象,静态成员 s 加 1
    id=s;　　　　　　//利用静态成员处理记录的编号,实现该类对象的记录编号自动递增
    name=name1;　　　//参数 name1 赋值给成员数据 name
    sex=sex1;　　　　//参数 sex1 赋值给成员数据 sex
    email=email1;　　//参数 email1 赋值给成员数据 email
    address=address1;//参数 address1 赋值给成员数据 address
    telnumber=tn;　　//参数 tn 赋值给成员数据 telnumber
    type=type1;　　　//参数 type1 赋值给成员数据 type
}
void CPhoneRecord::SetRecord(string name1,string sex1,string email1,string address1,string tn, int type1)
//用户函数形参为成员数据赋值
{
    name=name1;
    sex=sex1;
    email=email1;
    address=address1;
    telnumber=tn;
    type=type1;
}
string CPhoneRecord::GetName()
{
    return name;　　　　//返回姓名
}
string CPhoneRecord::GetSex()
{
    return sex;　　　　　//返回性别
}
string CPhoneRecord::GetEmail()
{
    return email;　　　　//返回邮箱
}
string CPhoneRecord::GetAddress()
{

```cpp
    return address;              //返回地址
}
string CPhoneRecord::GetTelnumber()
{
    return telnumber;            //返回电话号码
}
int CPhoneRecord::GetType()
{
    return type;                 //返回分组类别
}
void CPhoneRecord::SetTelNumber(string number)
{
    telnumber=number;            //设置电话号码
}
void CPhoneRecord::Display()    //显示信息
{
    cout<<setw(6)<<id<<setw(10)<<name
        <<setw(6)<<sex<<setw(15)<<email
        <<setw(15)<<address<<setw(18)<<telnumber
        <<setw(8)<<TypeString(type)<<endl;
}
string CPhoneRecord::TypeString(int type)        //将整型type转换成字符串分组
{
    string t;
    if(type==1)
        t="朋友";
    else if(type==2)
        t="同学";
    else if(type==3)
        t="家人";
    else if(type==4)
        t="同事";
    return t;
}
class CNode                                      //定义链表的结点类
{
private:
    CPhoneRecord * phoneRecord;                  //结点的数据域部分
```

```cpp
    CNode * pNext;                            //结点的指针域部分,指向下一个结点
public:
    CNode();
    CNode(CNode &node);                       //构造函数,对象的初始化
    void SetPhoneData(CPhoneRecord * pdata);  //为成员数据 phoneRecord 赋值
    void DisplayNode();                       //显示结点的数据域信息
    CPhoneRecord * GetData();                 //获取成员数据 phoneRecord 指向的数据
    friend class CList;                       //将 Clist 类定义为 CNode 类的友元类
};
CNode::CNode()
{
    phoneRecord=0;pNext=0;                    //将数据域指针设置为 NULL,将指针域设置为 NULL
}
CNode::CNode(CNode &node)
{
    phoneRecord=node.phoneRecord;
    //将参数 node 的 phoneRecord 赋值给成员数据 phoneRecord
    pNext=node.pNext;                         //将参数 node 的 pNext 赋值给成员数据 pNext
}
void CNode::SetPhoneData(CPhoneRecord * pdata)    //设置结点的数据域
{
    phoneRecord=pdata;
}
void CNode::DisplayNode()                     //显示结点的数据域
{
    phoneRecord->Display();
}
CPhoneRecord * CNode::GetData()               //获取结点的数据域
{
    return phoneRecord;
}
class CList                                   //定义链表类
{
    CNode * pHead;                            //链表的头指针
public:
    CList()
    {
        pHead=0;                              //将 pHead 赋值为 NULL
```

```cpp
    }
    ~CList()
    {
        DeleteList();                        //对象销毁前,释放链表所占的存储单元
    }
    void AddRecord();
    //增加记录,用于从文件中读取数据时建立链表
    void AddRecord(CNode * pnode);
    //增加记录,用于通过输入数据增加新结点
    void DeleteRecord();                     //删除记录
    void FindRecord();                       //根据姓名查找记录
    void FindRecordClass();                  //根据分组查找记录
    void UpdateRecord();                     //更新记录
    void DisplayList();                      //显示所有记录
    void DeleteList();                       //删除所有记录
    CNode * GetListHead(){return pHead;}     //获取链表的头结点
    CNode * GetListNextNode(CNode * pnode);  //获取下一个结点
};

CNode * CList::GetListNextNode(CNode * pnode)
{
    CNode * p1=pnode;
    return p1->pNext;
}
void CList::AddRecord()
{
    CNode * pNode;                           //定义指向 CNode 类对象的指针变量
    CPhoneRecord * pPhone;                   //定义 CPhoneRecord 类对象的指针变量
    string name,sex,email,address,telnumber;
    int type;
    cout<<"输入姓名(输入 0 结束):";
    cin>>name;
    while(name!="0")//循环重复添加数据,直到姓名输入值为"0"为止
    {
        cout<<"\n 输入性别:";
        cin>>sex;
        cout<<"\n 输入邮箱:";
        cin>>email;
```

```cpp
        cout<<"\n 输入住址:";
        cin>>address;
        cout<<"\n 输入电话号码:";
        cin>>telnumber;
        cout<<"\n 输入分组类型 1、朋友,2、同学,3、家人,4、同事:";
        cin>>type;
        pPhone=new CPhoneRecord;        //创建 CPhoneRecord 类对象
        pPhone->SetRecord(name,sex,email,address,telnumber,type);
        //为新创建 CPhoneRecord 类对象赋值
        pNode= new CNode;               //创建 CNode 类对象
        pNode->SetPhoneData(pPhone);    //设置结点的数据域
        AddRecord(pNode);               //将该结点添加到链表中
        cout<<"输入姓名(输入 0 结束):";
        cin>>name;
    }
    cout<<endl<<endl;
    system("pause");
}
void CList::AddRecord(CNode * pNode)
{
    if(pHead==0)                        //链表为空,则该结点为链表的头结点
    {
        pHead=pNode;
        pNode->pNext=0;                 //结点的指针域赋值为 NULL
        return ;
    }
    else                                //否则,插入到链表的首部
    {
        pNode->pNext=pHead;             //新结点 pNode 的指针域指向头结点
        pHead=pNode;                    //新的结点 pNode 作为头结点
    }
    cout<<endl<<endl;
}
void CList::DeleteRecord()
{
    CNode * p1,* p2;                    //定义指向 CNode 对象的指针变量 p1 和 p2
    char f;                             //定义字符变量 f
    string name;
```

```cpp
        cout<<"输入您需要删除的姓名(输入0结束)";
        cin>>name;
        while(name!="0")                    //循环重复删除数据,直到姓名输入值为"0"为止
        {
            p1=pHead;
            while(p1&&p1->GetData()->GetName()!=name)   //查找要删除的记录结点
            {
                p2=p1;                      //保持p1为正在查找的结点指针,p2为p1前一个结
                                            //  点的指针
                p1=p1->pNext;
            }
            if(p1==NULL)                    //找不到记录,给出提示
                cout<<"在通讯录中查找不到"<<name<<"."<<endl;
            else
            {
                cout<<"在通讯录中找到"<<name<<",内容是:"<<endl;
                p1->DisplayNode();          //显示结点数据
                cout<<"确定要删除"<<name<<"的资料吗,Y:N?";
                cin>>f;                     //用户输入是否确认删除
                if(f=='Y')                  //确认删除
                {
                    if(p1==pHead)           //删除的结点为头结点
                    {
                        pHead=pHead->pNext;
                        delete p1;          //释放p1指向的存储单元
                    }
                    else                    //删除的结点为非头结点
                    {
                        p2->pNext=p1->pNext;
                        delete p1;
                    }
                }
            }
            cout<<"输入您需要删除的姓名(输入0结束)";
            cin>>name;
        }
    system("pause");
}
```

```cpp
void CList::FindRecord()
{
    CNode *p1;                          //定义指向 CNode 对象的指针变量 p1
    string name;
    cout<<"输入您需要查找的姓名(输入 0 结束)";
    cin>>name;
    while(name!="0")                    //循环重复查找数据,直到姓名输入值为"0"为止
    {
        p1=pHead;                       //将头结点指针 pHead 赋值给 p1
        while(p1)                       //查找结点
        {
            if(p1->GetData()->GetName()==name)
                break;                  //找到符合要求的结点,则退出循环
            p1=p1->pNext;
        }
        if(p1)                          //p1 不为 NULL
        {
            cout<<"在通讯录中找到"<<name<<",内容是:"<<endl;
            p1->DisplayNode();          //显示结点信息
        }
        else
            cout<<"在通讯录中查找不到啊"<<name<<". "<<endl;
        cout<<"输入您需要查找的姓名(输入 0 结束)";
        cin>>name;
    }
    cout<<endl<<endl;
    system("pause");
}
void CList::FindRecordClass()
{
    CNode *p1;                          //定义指向 CNode 对象的指针变量 p1
    int type;
    cout<<"输入您需要查找的分组记录(输入 0 结束)";
    cout<<"\n 分组类型 1、朋友,2、同学,3、家人,4、同事:";
    cin>>type;
    int f;
    while(type!=0)                      //循环重复查找数据,直到分组类别输入值为 0 为止
    {
```

```cpp
            f=0;
            p1=pHead;                        //将 p1 赋值为头结点指针
            while(p1)                        //查找结点
            {
                if(p1->GetData()->GetType()==type)
                {
                    p1->DisplayNode();       //找到符合要求的结点,则退出循环
                    f=1;
                }
                p1=p1->pNext;
            }
            cout<<endl<<endl;
            if(f==0)
                cout<<"在通讯录中查找不到该分组的记录啊"<<"。"<<endl;
            cout<<"输入您需要查找的分组记录(输入 0 结束)";
            cout<<"\n 分组类型 1,朋友,2,同学,3,家人,4,同事:";
            cin>>type;
    }
    cout<<endl<<endl;
    system("pause");
}
void CList::DisplayList()
{
    CNode *p1=pHead;
    //定义指向 CNode 对象的指针变量 p1 并初始化为头结点指针
    cout<<setw(6)<<"编号"<<setw(10)<<"姓名"
        <<setw(6)<<"性别"<<setw(15)<<"email"
        <<setw(15)<<"家庭住址"<<setw(18)<<"电话号码"
        <<setw(8)<<"分组"<<endl;
    while(p1)//从链表头结点开始输出链表的每一个的结点直到 p1 为 NULL 为止
    {
        p1->phoneRecord->Display();
        p1=p1->pNext;
    }
    cout<<endl<<endl;
    system("pause");
}
void CList::DeleteList()
```

```
    {
        CNode *p1,*p2;              //定义指向 CNode 对象的指针变量 p1 和 p2
        p1=pHead;                   //将 p1 赋值为头结点指针
        while(p1)
        {
            delete p1->phoneRecord;
            //释放结点的数据域 phoneRecord 所指向的存储单元
            p2=p1;
            p1=p1->pNext;
            delete p2;              //释放 p2 所指向的存储单元
        }
    }
void CList::UpdateRecord()
{
    CNode *p1;                      //定义指向 CNode 对象的指针变量 p1
    string name,sex,email,address,telnumber;
    int type;
    cout<<"输入您需要修改电话的姓名(输入 0 结束)";
    cin>>name;
    while(name!="0")                //循环重复修改数据,直到姓名输入值为"0"为止
    {
        p1=pHead;
        while(p1)                   //查找要修改的结点
        {
            if(p1->GetData()->GetName()==name)
                break;              //找到需要修改的结点,则退出循环
            p1=p1->pNext;
        }
        if(p1)                      //p1 不为 NULL,即找到符合要求结点
        {
            cout<<"在通讯录中找到"<<name<<",内容是:"<<endl;
            p1->DisplayNode();      //显示记录修改前的信息
            cout<<"\n 请输入新的性别:";      cin>>sex;
            cout<<"\n 请输入新的邮箱:";      cin>>email;
            cout<<"\n 请输入新的住址:";      cin>>address;
            cout<<"\n 请输入新的电话号码:";  cin>>telnumber;
            cout<<"\n 输入新的分组类型 1、朋友,2、同学,3、家人,4、同事:";
            cin>>type;
```

```cpp
            p1->GetData()->SetRecord(name,sex,email,address,telnumber,type);
            cout<<"通讯录中"<<name<<",新的内容是:"<<endl;
            p1->DisplayNode();              //显示记录修改后的信息
        }
        else                                //找不到结点
            cout<<"在通讯录中查找不到啊"<<name<<"."<<endl;
        cout<<"输入您需要查找的姓名(输入0结束)";
        cin>>name;
    }
    cout<<endl<<endl;
    system("pause");
}
void StoreFile(CList &PhoneList)                //将数据写入到文件中
{
    ofstream outfile("TELEPHONEW.DAT",ios::out);//定义输出文件对象
    if(! outfile)                               //文件打开错误
    {
        cout<<"数据文件打开错误,没有将数据存入文件!\n";
        return;
    }
    CNode * pnode;
    CPhoneRecord * pPhone;
    pnode=PhoneList.GetListHead();              //将链表头结点指针赋值给指针变量pnode
    while(pnode)
    {
        pPhone=(CPhoneRecord *)pnode->GetData();//获取结点的数据域指针
        outfile<<endl;
        outfile<<pPhone->GetName()<<"\t"<<pPhone->GetSex()<<"\t";
        outfile<<pPhone->GetEmail()<<"\t"<<pPhone->GetTelnumber()<<"\t";
        outfile<<pPhone->GetAddress()<<"\t"<<pPhone->GetType();
        //将当前结点的数据信息写入到文件中
        pnode=PhoneList.GetListNextNode(pnode); //获取下一个结点指针
    }
    outfile.close();                            //关闭文件对象
}
void Operate(string &strChoice,CList &PhoneList)
{
    if(strChoice=="1")
```

```
            PhoneList.AddRecord();                 //增加通讯录记录
        else if(strChoice=="2")
            PhoneList.DisplayList();               //显示所有通讯录记录
        else if(strChoice=="3")
            PhoneList.FindRecord();                //查询通讯录记录
        else if(strChoice=="4")
            PhoneList.FindRecordClass();           //查询通讯录记录
        else if(strChoice=="5")
            PhoneList.DeleteRecord();              //删除通讯录记录
        else if(strChoice=="6")
            PhoneList.UpdateRecord();              //修改电话簿记录
        else if(strChoice=="0")
            StoreFile(PhoneList);                  //保存数据到文件中
        else
            cout<<"输入错误,请重新输入您的选择:";
}
void LoadFile(CList &PhoneList)
{
    ifstream infile("TELEPHONEW.DAT",ios::in);    //定义输入文件对象
    if(!infile)                                    //文件打开错误
    {
        cout<<"没有数据文件!!! \n\n";
        return ;
    }
    string name,email,sex,address,telephone;
    CNode *pNode;
    CPhoneRecord *pPhone;
    int type;
    while(!infile.eof())     //从文件中读数据,直到文件指针指向文件末尾为止
    {
        infile>>name>>sex>>email>>telephone>>address>>type;
        //从文件中读取数据给相应变量
        pPhone=new CPhoneRecord(name,sex,email,address,telephone,type);
        //利用相应的值创建 CPhoneRecord 对象
        pNode=new CNode;                          //创建新的 CNode 结点对象
        pNode->SetPhoneData(pPhone);              //设置结点的数据域
        PhoneList.AddRecord(pNode);               //将新结点插入到链表中
    }
```

```cpp
        infile.close();                                 //关闭文件
}
int CPhoneRecord::s=0;                                  //设置静态成员变量s为0
int main(void)
{
        CList PhoneList;                                //定义链表对象PhoneList
        system("cls");                                  //清屏
        cout<<"\t欢迎进入通讯录数据系统\n";
        LoadFile(PhoneList);                            //从文件中加载数据,生成链表
        string strChoice;
        do
        {
                cout<<"\t1.添加通讯录记录\n";
                cout<<"\t2.显示通讯录记录\n";
                cout<<"\t3.根据姓名查询通讯录记录\n";
                cout<<"\t4.根据分组查询通讯录记录\n";
                cout<<"\t5.根据姓名删除通讯录记录\n";
                cout<<"\t6.根据姓名修改通讯录记录\n";
                cout<<"\t0.退出系统\n\n\n";
                cout<<"\n请输入您的选择:";
                //显示菜单
                cin>>strChoice;                         //输入用户的选项
                Operate(strChoice,PhoneList);           //调用Operate函数处理用户的选择
        }while(strChoice!="0");                         //输入"0",退出系统
        StoreFile(PhoneList);                           //保存数据到文件中
        cout<<"\n\n欢迎再次使用通讯录数据系统\n\n";
        return 0;
}
```

样例三 学生选课系统

一、设计的任务要求

设计一个简易的学生选课系统,能够完成课程信息的增加、删除和查找,学生信息的增加、删除和查找,学生选课,学生取消选课等操作,数据信息使用文件保存。要求系统具有菜单和提示,界面友好。

二、程序功能设计

学生选课系统涉及学生、课程、教师三个对象,学生和教师两种角色,为简化系统本系统对于教师只设计了一个管理员教师。学生具有查询本人信息,查询课程信息,选课和取消选课的等功能;教师具有课程信息的添加、删除、查询,学生信息的添加、删除和查询,选课、取消选课等功能。现将选课功能细化,结构如图 2.3.1 所示

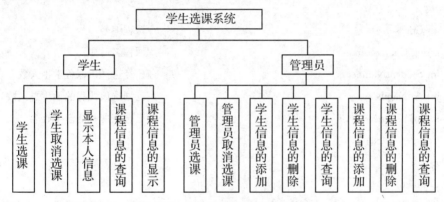

图 2.3.1 学生选课系统功能结构图

学生信息的添加:添加学生信息,按照学生的学号由小到大存放学生信息。

学生信息的删除:删除学生信息,根据学生的学号删除学生信息。

学生信息的查找:查询学生信息,根据学生的学号或姓名查询学生信息。在学生登录时以学号和姓名作为用户的编号和密码,即也提供了根据学号和姓名查询学生信息。

课程信息的添加:添加课程信息,按照课程的编号由小到大存放课程信息。

课程信息的删除:删除课程信息,按照课程的编号删除课程信息。

课程信息的查找:查找课程信息,根据课程的编号和名称查询课程的信息。

选课:选课可以由学生选,也可以由管理员进行选课。若想要成功选课,需要满足两个条件,一是学生选修课程未满并且没有选修该门课程,二是课程选修人数没有达到上限。当满足选课条件时,将课程名添加到学生已选课程中,将学生已选课程数加1,将学生姓名添加

到课程的已选学生中,将课程的已选学生人数加 1。

取消选课:取消选课可以由学生自己取消,也可以由管理员取消。操作和选课时相反,不作详细说明。

存储数据:将学生信息和课程信息以文件的形式进行存储。

载入数据:将学生信息和课程信息从文件中读出。

登录:具有学生和管理员两种角色。为简化系统,学生登录时以学号和姓名进行登录,管理员只有一个密码设置也比较简单。

三、程序设计的完整源代码以及注解

```cpp
#include<iostream>
#include<string>
#include<fstream>
using namespace std;
class Course                                          //定义课程类
{
private:
    int CourseId;                                     //课程编号
    string CourseName;                                //课程名
    int StudentNumber;                                //已经选择该课程学生人数
    int AllNumber;                                    //计划选该课程的学生总人数
    string SName[80];                                 //已选该课程的学生姓名
    Course * next;
public:
    Course(){}
    Course(int ,string,int ,int,Course * );           //Course 的构造函数,完成初始化工作
    void SetNext(Course * p);                         //设置该课程的下一课程
    Course * GetNext();                               //获取该课程的下一课程指针
    void ShowCourse();                                //显示课程的所有信息
    int GetId();                                      //获取课程编号
    string GetName();                                 //获取课程名称
    bool permit();                                    //是否允许学生选课
    void SelectCourse(string);                        //学生选课时课程对象的更改
    void CancelCourse(string);                        //学生取消选课时课程对象的更改
    friend void SaveCourse(Course * p);
    //将存储课程文件函数 SaveCourse 作为该类的友元函数
    friend Course *   LoadCourse();
```

//将读取课程文件函数 LoadCourse 作为该类的友元函数
};
Course::Course(int ci,string n,int an,int s,Course * p)
{
    CourseId=ci;
    CourseName=n;
    StudentNumber=s;
    AllNumber=an;
    next=p;
}
void Course::ShowCourse()   //显示课程信息
{
    cout<<CourseId<<"\t"<<CourseName;
    cout<<"\t"<<AllNumber<<"\t\t"<<StudentNumber<<"\t\t";
    for(int i=0;i<StudentNumber;i++)   //显示已经选择该课程的学生姓名
    {
        cout<<"\t"<<SName[i];
    }
}
Course *   Course::GetNext()
{
    return next;   //返回下一结点指针
}
void Course::SetNext(Course * p)
{
    next=p;   //设置 next 域
}
int Course::GetId()
{
    return CourseId;   //获取课程的 ID
}
string Course::GetName()
{
    return CourseName;   //获取课程名
}
bool Course::permit()

```cpp
{
    if(StudentNumber>=AllNumber)         //判断该课程是否可选
    {
        cout<<CourseName<<"课程,选课人数已满!!!"<<endl;
        return false;                    //不可选返回 false
    }
    else
        return true;                     //可选返回 true
}
void Course::SelectCourse(string name)
{
    int i=0;
    for(;i<StudentNumber;i++)            //在选课学生里查找该学生,找到终止循环
    {
        if(SName[i]==name)
            break;
    }
    if(i>=StudentNumber)                 //学生 name 没有选择该课程,进行选课处理
    {
        SName[i]=name;
        StudentNumber+=1;
        cout<<"你已经选择了"<<CourseName<<"课程。"<<endl;
    }
    else                                 //学生 name 已经选择了该课程
        cout<<"不好意思,同学你已经选择"<<CourseName<<"课程,请不要重复选择哦!!!"<<endl;
}
void Course::CancelCourse(string name)
{
    for(int i=0;i<StudentNumber;i++)     //在选课学生里查找该学生,找到终止循环
        if(SName[i]==name)
            break;
    if(i>=StudentNumber)                 //学生 name 没有选择该课程
        cout<<"不好意思,同学你没有选择"<<CourseName
            <<"课程,没办法取消选课哦!!!"<<endl;
    else                                 //学生 name 已经选择了该课程,进行取消选课处理
```

```cpp
        {
            for(;i<StudentNumber-1;i++)      //将该学生从该课程的选课学生里删除
                SName[i]=SName[i+1];
            StudentNumber-=1;                //将已选学生人数减1
            cout<<"你已经取消了"<<CourseName<<"的课程选课"<<endl;
        }
}
class Student                                //学生信息类
{
private:
    int StudentId;                           //学生学号
    string StudentName;                      //学生姓名
    int AllNumber;                           //可选课程数
    int CourseNumber;                        //已选课程数
    string CName[3];                         //已选课程名称
    Student * next;                          //指向下一个学生指针
public:
    Student(){}
    Student(int ,string,int ,int,Student * );   //初始化数据
    void SetNext(Student * );                //设置指向下一学生的指针
    Student * GetNext();                     //获取指向下一学生的指针
    int GetSId();                            //获取学生学号
    string GetSName();                       //获取学生姓名
    bool permit();
    void SelectCourse(string);               //学生选课时学生信息的更改
    void CancelCourse(string);               //学生取消选课时学生信息的更改
    void ShowStudent();                      //显示学生信息
    friend void SaveStudent(Student * p);    //存储学生文件函数作为该类的友元函数
    friend Student * LoadStudent();          //加载学生文件函数作为该类的友元函数
};
Student::Student(int i ,string n,int a ,int b,Student * p)
{
    StudentId=i;
    StudentName=n;
    AllNumber=a;
    CourseNumber=b;
    next=NULL;
```

```cpp
        cout<<i<<endl;
}
void Student::SetNext(Student * ps)
{
    next=ps;                        //设置next域
}

Student * Student::GetNext()
{
    return next;                    //返回next
}
int Student::GetSId()
{
    return StudentId;               //返回学生学号
}
string Student::GetSName()
{
    return StudentName;             //返回学生姓名
}
bool Student::permit()
{
    if(CourseNumber>=AllNumber)     //判断学生是否已经选满课程
    {
        cout<<"你的课程已经选满了,不要多选哦!!!"<<endl;
        return false;//允许选课返回false
    }
    else
        return true;//不允许选课返回true
}
void Student::SelectCourse(string n)
{

    for(int i=0;i<CourseNumber;i++) //在学生已选课程里查找课程名为n的课程
        if(CName[i]==n)
            break;
    if(i<CourseNumber)              //学生已经选择了该课程
```

```cpp
        cout<<"同学,你已经选了课程"<<n<<",不可以重复选哦!!!"<<endl;
    else                            //学生没有选择该课程,进行学生选课处理
    {
        CName[CourseNumber]=n;      //将新课程加入该学生已选课程名
        CourseNumber+=1;            //学生已选课程数加1
    }
}
void Student::CancelCourse(string n)
{
    for(int i=0;i<CourseNumber;i++)  //在学生已选课程里查找课程名为n的课程
        if(CName[i]==n)
            break;
    if(i>=CourseNumber)              //学生没有选择该课程
        cout<<"同学,你没有选择"<<n<<"课程,不可以取消哦!!!!!"<<endl;
    else                             //学生已经选择了该课程,进行取消选课处理
    {
        for(;i<CourseNumber-1;i++)   //将课程从已选课程名中删除
            CName[i]=CName[i+1];
        CourseNumber-=1;             //学生已选课程数减1
        cout<<"你已经取消了"<<n<<"的课程选课。"<<endl;
    }
}
void Student::ShowStudent()          //显示学生信息
{
    cout<<StudentId<<"\t"<<StudentName;
    cout<<"\t"<<AllNumber<<"\t"<<CourseNumber;
    for(int i=0;i<CourseNumber;i++)  //显示学生已经选择课程的课程名
    {
        cout<<"\t"<<CName[i];
    }
}
class ElectiveCourse                 //学生选课类
{
private:
    Course * pCourse;                //指向课程链表的指针
    Student * pStudent;              //指向学生链表的指针
public:
```

```cpp
    ElectiveCourse(){pCourse=NULL;pStudent=NULL;}
    ElectiveCourse(Course *,Student *);       //初始化数据
    void AddCourse();                          //增加课程
    void SubCourse();                          //删除课程
    void AddStudent();                         //增加学生
    void SubStudent();                         //删除学生
    void FindCourseId();                       //根据课程编号查询课程信息
    Course * FindCourseName(string);           //根据课程名称查询课程信息
    void FindCourseName();                     //根据课程名称查询课程信息
    void FindStudentId(int);                   //根据学生学号查询学生信息
    Student * FindStudentName(string);         //根据学生姓名查询学生信息
    void FindStudentName();                    //根据学生姓名查询学生信息
    bool FindStudent(int ,string);             //根据学生学号和姓名查询学生信息,用于密码验证
    void StSeCourse(string);                   //学生自己选课
    void MaSeCourse();                         //管理员帮学生选课
    void StCaCourse(string);                   //学生自己取消选课
    void MaCaCourse();                         //管理员帮学生取消选课
    void ShowCourse();                         //显示所有课程信息
    void ShowStudent();                        //显示所有学生信息
    Course * GetCourse();                      //返回课程链表的头指针
    Student * GetStudent();                    //返回学生链表的头指针
};
ElectiveCourse::ElectiveCourse(Course * pc,Student * ps)     //构造函数,初始化对象
{
    pCourse=pc;
    pStudent=ps;
}
void ElectiveCourse::AddCourse()               //增加课程
{
    Course *p1,*p2,*p3;                        //定义指向Course类的对象的指针变量p1、p2和p3
    p1=pCourse;                                //将课程链表的头结点指针赋值给p1
    int id;
    string name;
    int an;
    cout<<"\n请输入课程编号:";
    cin>>id;
    cout<<"\n请输入课程名称:";
```

```cpp
    cin>>name;
    cout<<"\n请输入该课程计划学生数(不要超过 80):";
    cin>>an;
    p3=new Course(id,name,an,0,0);        //创建 Course 对象并利用输入数据初始化
    if(pCourse==NULL)//链表为空,将新结点作为链表的头结点
    {
        pCourse=p3;
        cout<<"你已经成功的添加了"<<name<<"课程!!"<<endl;
        return ;
    }
    while(p1&&p1->GetId()<id)
    //查找结点的课程编号大于等于要插入结点课程编号的第一个结点
    //指针 p1 表示符合条件的结点的指针,指针 p2 是指针 p1 的前一个结点指针
    {
        p2=p1;
        p1=p1->GetNext();
    }
    if(p1==pCourse)                    //插入位置为头结点前
    {
        p1->SetNext(pCourse);
        pCourse=p1;
    }
    else                               //插入位置为链表的中间和链表尾部
    {
        p2->SetNext(p3);
        p3->SetNext(p1);
    }
    cout<<"你已经成功的添加了"<<name<<"课程!!"<<endl;
}
void ElectiveCourse::SubCourse()
{
    Course *p1,*p2;                    //定义指向 Course 类的对象的指针变量 p1 和 p2
    int id;
    cout<<"请输入需要删除的课程的编号:";
    cin>>id;
    cout<<endl;
    p1=pCourse;                        //将课程链表的头结点指针赋值给 p1
```

```cpp
    if(pCourse==NULL)                    //链表为空
    {
        cout<<"表中没有任何课程,你不能删除课程!!!!!"<<endl;
        return;
    }
    while(p1&&p1->GetId()!=id)
    //查找结点的课程号等于要删除课程号的第一个结点
    //指针 p1 表示符合条件的结点的指针,指针 p2 是指针 p1 的前一个结点指针
    {
        p2=p1;
        p1=p1->GetNext();
    }
    if(p1==NULL)                         //没有找到符合要求的结点
    {
        cout<<"没有找到课程编号为"<<id<<"的课程,不可以删除课程!!!!"<<endl;
    }
    else
    {
        if(p1==pCourse)                  //删除的结点为头结点
        {
            pCourse=p1->GetNext();
            delete p1;                   //释放 p1 所指向的结点
        }
        Else                             //删除的结点为非头结点
        {
            p2->SetNext(p1->GetNext());
            delete p1;
        }
    }
}
void ElectiveCourse::AddStudent()
{
    Student *p1,*p2,*p3;                 //定义指向 Student 类的对象的指针变量 p1、p2 和 p3
    p1=pStudent;                         //将学生链表的头结点指针赋值给 p1
    int id;
    string name;
    cout<<"\n 请输入学生学号:";
```

```cpp
        cin>>id;
        cout<<"\n请输入学生姓名:";
        cin>>name;
        p3=new Student(id,name,3,0,0);        //创建新的Student对象,并初始化
        if(pStudent==NULL)//链表为空
        {
            pStudent=p3;
            cout<<"你已经成功的添加了"<<name<<"学生!!"<<endl;
            return;
        }
        while(p1&&p1->GetSId()<id)
        //查找结点的学号大于等于要插入结点学号的第一个结点
        //指针p1表示符合条件的结点的指针,指针p2是指针p1的前一个结点指针
        {
            p2=p1;
            p1=p1->GetNext();
        }
        if(p1==pStudent)                      //插入位置为头结点前
        {
            p1->SetNext(pStudent);
            pStudent=p1;
        }
        else                                  //插入位置为链表的中间和链表尾部
        {
            p2->SetNext(p3);
            p3->SetNext(p1);
        }
        cout<<"你已经成功的添加了"<<name<<"学生!!"<<endl;
}
void ElectiveCourse::SubStudent()
{
    Student *p1,*p2;                          //定义指向Student类的对象的指针变量p1和p2
    int id;
    cout<<"请输入需要删除的学生的学号:";
    cin>>id;
    cout<<endl;
    p1=pStudent;                              //将学生链表的头结点指针赋值给p1
```

```cpp
    if(pStudent==NULL)              //链表为空
    {
        cout<<"表中没有任何学生,你不能删除学生!!!!!"<<endl;
        return;
    }
    while(p1&&p1->GetSId()!=id)
    //查找结点的学号等于要删除学生学号的第一个结点
    //指针 p1 表示符合条件的结点的指针,指针 p2 是指针 p1 的前一个结点指针
    {
        p2=p1;
        p1=p1->GetNext();
    }
    if(p1==NULL)                    //没有找到学生
    {
        cout<<"没有找到学生学号为"<<id<<"的学号,不可以删除学生!!!!"<<endl;
    }
    else
    {
        if(p1==pStudent)            //删除结点为头结点
        {
            pStudent=p1->GetNext();
            delete p1;              //释放 p1 所指向的结点
        }
        Else                        //删除结点为非头结点
        {
            p2->SetNext(p1->GetNext());
            delete p1;
        }
    }
}
void ElectiveCourse::FindCourseId()
{
    Course *p1;                     //定义指向 Course 类的对象的指针变量 p1
    int id;
    cout<<"请输入需要查找的课程的编号:";
    cin>>id;
    cout<<endl;
```

```
        p1=pCourse;                          //将课程链表的头结点指针赋值给 p1
        while(p1&&p1->GetId()!=id)
        //查找结点的课程号等于要删除课程号的第一个结点
        //指针 p1 表示符合条件的结点的指针,指针 p2 是指针 p1 的前一个结点指针
        {
                p1=p1->GetNext();
        }
        if(p1==NULL)                         //没有找到符合要求的结点
                cout<<"没有找到课程编号为"<<id<<"的课程。"<<endl;
        else
                p1->ShowCourse();            //找到显示结点信息
}
Course * ElectiveCourse::FindCourseName(string name)
{
        Course *p1;                          //定义指向 Course 类的对象的指针变量 p1
        p1=pCourse;                          //将课程链表的头结点指针赋值给 p1
        while(p1&&p1->GetName()!=name)
        //查找结点的课程名称等于要删除课程名称的第一个结点
        //指针 p1 表示正在查找结点的指针
        {
                p1=p1->GetNext();
        }
        return p1;
}
void ElectiveCourse::FindCourseName()
{
        Course *p1;                          //定义指向 Course 类的对象的指针变量 p1
        string name;
        cout<<"请输入需要查找的课程的名称:";
        cin>>name;
        cout<<endl;
        p1=pCourse;                          //将课程链表的头结点指针赋值给 p1
        while(p1&&p1->GetName()!=name)
        //在链表中查找结点的课程名称等于要删除课程名称的第一个结点
        //指针 p1 表示正在查找结点的指针
        {
                p1=p1->GetNext();
```

```cpp
    }
    if(p1==NULL)                    //没有找到符合要求的结点
        cout<<"没有找到课程名为"<<name<<"的课程。"<<endl;
    else
        p1->ShowCourse();           //找到并显示结点信息
}
void ElectiveCourse::FindStudentId(int id)
{
    Student *p1;                    //定义指向Student类的对象的指针变量p1
    p1=pStudent;                    //将学生链表的头结点指针赋值给p1
    while(p1&&p1->GetSId()!=id)
    //在链表中查找结点的学号等于要删除学生学号的第一个结点
    //指针p1表示正在查找的结点指针
    {
        p1=p1->GetNext();
    }
    if(p1==NULL)                    //没有找到学生
        cout<<"没有找到学生学号为"<<id<<"的课程！！！！"<<endl;
    else
        p1->ShowStudent();          //找到结点，并显示结点信息
}
Student * ElectiveCourse::FindStudentName(string name)
{
    Student *p1;                    //定义指向Student类的对象的指针变量p1
    p1=pStudent;                    //将学生链表的头结点指针赋值给p1
    while(p1&&p1->GetSName()!=name)
    //在链表中查找结点的学生姓名等于要删除学生姓名的第一个结点
    //指针p1表示正在查找的结点的指针
    {
        p1=p1->GetNext();
    }
    return p1;                      //返回找到的结点指针
}
bool ElectiveCourse::FindStudent(int id,string name)
//根据学号和姓名查找学生信息，用于学生用户登录
{
    Student *p1;
```

```cpp
        p1=pStudent;
        while(p1)
        {
            if(p1->GetSId()==id&&p1->GetSName()==name)
                break;
            p1=p1->GetNext();
        }
        if(p1)
            return true;
        else
            return false;
}
void ElectiveCourse::FindStudentName()
{
        Student * p1;
        string name;
        cout<<"请输入需要查找的学生的姓名:";
        cin>>name;
        cout<<endl;
        p1=pStudent;
        while(p1&&p1->GetSName()!=name)
        {
            p1=p1->GetNext();
        }
        if(p1==NULL)
            cout<<"没有找到学生姓名为"<<name<<"的学生!!!!"<<endl;
        else
            p1->ShowStudent();
}
void ElectiveCourse::MaSeCourse()          //管理员帮学生选课
{
        string sname,cname;
        cout<<"\n请输入选课学生姓名:";
        cin>>sname;
        Student * p=FindStudentName(sname);//查找学生姓名为 sname 的学生
        cout<<"\n请输入学生选择课程名:";
        cin>>cname;
```

```cpp
    Course * pc=FindCourseName(cname);         //查找课程名为 cname 的课程
    if(p==NULL)                                 //学生不存在
    {
        cout<<"该学生不存在!!!";
        return ;
    }
    if(pc==NULL)                                //课程不存在
    {
        cout<<"该课程不存在";
        return ;
    }
    if(p->permit()&&pc->permit())               //允许选课
    {
        pc->SelectCourse(sname);                //选课时处理课程信息的更改
        p->SelectCourse(cname);                 //选课时处理学生信息的更改
    }
}
void ElectiveCourse::StSeCourse(string sname)   //学生登录自己选课
{
    string cname;
    cout<<"\n请输入学生选择课程名:";
    cin>>cname;
    Student * p=FindStudentName(sname);         //查找学生姓名为 sname 的学生
    Course * pc=FindCourseName(cname);          //查找课程名为 cname 的课程
    if(pc==NULL)                                //课程不存在
    {
        cout<<"该课程不存在";
        return ;
    }
    if(p->permit()&&pc->permit())               //允许选课
    {
        pc->SelectCourse(sname);                //选课时处理课程信息的更改
        p->SelectCourse(cname);                 //选课时处理学生信息的更改
    }
}
void ElectiveCourse::StCaCourse(string sname)   //学生登录,取消选课
{
```

```cpp
        string cname;
        cout<<"\n 请输入学生选择课程名:";
        cin>>cname;
        Student * p=FindStudentName(sname);        //查找学生姓名为 sname 的学生
        Course * pc=FindCourseName(cname);         //查找课程名为 cname 的课程
        if(pc==NULL)                               //课程不存在
        {
            cout<<"该课程不存在";
            return ;
        }
        p->CancelCourse(cname);                    //取消选课时处理学生信息的更改
        pc->CancelCourse(sname);                   //取消选课时处理课程信息的更改
}
void ElectiveCourse::MaCaCourse()                  //管理员登录,帮学生取消选课
{
        string sname,cname;
        cout<<"\n 请输入选课学生姓名:";
        cin>>sname;
        Student * p=FindStudentName(sname);        //查找学生姓名为 sname 的学生
        cout<<"\n 请输入学生选择课程名:";
        cin>>cname;
        Course * pc=FindCourseName(cname);         //查找课程名为 cname 的课程
        if(p==NULL)                                //学生不存在
        {
            cout<<"该学生不存在!!!";
            return ;
        }
        if(pc==NULL)                               //课程不存在
        {
            cout<<"该课程不存在";
            return ;
        }
        p->CancelCourse(cname);                    //取消选课时处理学生信息的更改
        pc->CancelCourse(sname);                   //取消选课时处理课程信息的更改
}
Course * ElectiveCourse::GetCourse()
{
```

```cpp
    return pCourse;                          //获取课程链表的头结点指针
}
Student * ElectiveCourse::GetStudent()
{
    return pStudent;                         //获取学生链表的头结点指针
}
void ElectiveCourse::ShowCourse()
{
    Course * p=pCourse;                      //指向课程链表的指针
    cout<<"课程号"<<<<"\t"<<"课程名"<<<<"\t"
        <<"计划人数"<<<<"\t"<<"实际人数"<<<<"\t"<<"已选学生姓名";
    while(p)
    {
        cout<<endl;
        p->ShowCourse();                     //显示结点信息
        p=p->GetNext();                      //将p指针移到下一结点
    }
}
void ElectiveCourse::ShowStudent()           //显示所有学生的选课信息
{
    Student * p=pStudent;
    cout<<"学号"<<<<"\t 姓名"<<<<"\t 计划课程数"<<<<"\t 已选课程数"<<<<"\t 已选课程名";
    while(p)
    {cout<<endl;
        p->ShowStudent();
        p=p->GetNext();
        cout<<endl;
    }
}
void SaveCourse(Course * p)                  //存储课程数据到文件
{
    ofstream ofile;                          //定义输出文件对象
    ofile.open("Course.dat",ios::out);
    //以写的方式打开文件Course.dat,若该文件不存在,则创建Course.dat文件
    if(!ofile)                               //文件打开错误
    {
        cout<<"\n 数据文件打开错误! \n";
```

```cpp
        return ;
    }
    Course *t;
    while(p)
    {
        ofile<<endl;
        ofile<<p->CourseId<<"\t"<<p->CourseName<<"\t"<<p->AllNumber
            <<"\t"<<p->StudentNumber;
        for(int i=0;i<p->StudentNumber;i++)
            ofile<<"\t"<<p->SName[i];
        //将当前结点的数据信息写入到文件中
        t=p;p=p->next;
        delete t;                              //删除指针 t 所指向的结点
    }
    ofile.close();                             //关闭文件对象
}
Course *   LoadCourse()                        //加载课程文件
{
    ifstream ifile;                            //定义输入文件对象
    ifile.open("Course.dat",ios::in);          //以读的方式打开文件 Course.dat
    Course *p,*q,*h=NULL;
    if(!ifile)                                 //文件打开错误
    {
        cout<<"\n 数据文件不存在,加载不成功! \n";
        return NULL;
    }
    while(!ifile.eof())
    {
        p=new Course;                          //创建新的 Cours 对象
        ifile>>p->CourseId>>p->CourseName>>p->AllNumber>>p->StudentNumber;
        for(int i=0;i<p->StudentNumber;i++)
            ifile>>p->SName[i];
        //将数据从文件中读取到新的结点中
        p->next=NULL;
        if(h==NULL)
            q=h=p;
```

```cpp
            else
            {
                q->next=p;
                q=p;
            }                                       //创建链表
    }
    ifile.close();                                  //关闭文件对象
    return h;
}
void SaveStudent(Student * p)                       //存储学生数据到文件
{
    ofstream ofile;
    ofile.open("Student.dat",ios::out);
    if(!ofile)
    {
        cout<<"\n数据文件打开错误！\n";
        return ;
    }
    Student * t;
    while(p)
    {   ofile<<endl;
ofile<<p->StudentId<<"\t"<<p->StudentName<<"\t"<<p->AllNumber<<"\t"<<p->CourseNumber;
        for(int i=0;i<p->CourseNumber;i++)
            ofile<<"\t"<<p->CName[i];
        t=p;p=p->next;
        delete t;
    }
    ofile.close();
}
Student * LoadStudent()                             //加载学生文件
{
    ifstream ifile;
    ifile.open("Student.dat",ios::in);
    Student * p,* q,* h=NULL;
    if(!ifile)
    {
```

```cpp
        cout<<"\n 数据文件不存在,加载不成功! \n";
        return NULL;
    }
    while(!ifile.eof())
    {
        p=new Student;
        ifile>>p->StudentId>>p->StudentName>>p->AllNumber>>p->CourseNumber;
        for(int i=0;i<p->CourseNumber;i++)
            ifile>>p->CName[i];
        p->next=NULL;
        if(h==NULL)
            q=h=p;
        else
        {
            q->next=p;
            q=p;
        }
    }
    ifile.close();
    return h;
}
class Menu                                   //菜单的基类
{
public:
    Menu(){};
    void Show(){};                           //显示菜单
    char Get();                              //选择菜单
};
char Menu::Get()
{
    char ch;
    cin>>ch;
    return ch;
}
class SystemMenu:public Menu                 //系统菜单类
{
```

```cpp
public:
    SystemMenu(){}
    void Show();                          //显示系统菜单
};
void SystemMenu::Show()
{
    cout<<"\n功能菜单:"<<endl;
    cout<<"###############################\n";
    cout<<"\t1.管理员登录(A/a)"<<endl;
    cout<<"\t2.学生登录(S/s)"<<endl;
    cout<<"\t3.退出系统(Q/q)"<<endl;
    cout<<"###############################\n";
    cout<<"请输入选择的菜单:";
}
class AdMenu : public Menu                //管理员菜单类
{
public:
    AdMenu(){}
    void Show();                          //显示管理员菜单
};
void AdMenu::Show()
{
    cout<<"\n功能菜单:"<<endl;
    cout<<"* * * * * * * * * * * * * * * * * * * * * * *\n";
    cout<<"\t1.选课(1)"<<endl;
    cout<<"\t2.取消选课(2)"<<endl;
    cout<<"\t3.学生信息操作(3)"<<endl;
    cout<<"\t4.课程信息操作(4)"<<endl;
    cout<<"\t5.退出菜单(5)"<<endl;
    cout<<"* * * * * * * * * * * * * * * * * * * * * * *\n";
    cout<<"请输入选择的菜单:";
}
class StuMenu :public Menu                //学生菜单类
{
public:
    StuMenu(){}
    void Show();                          //显示学生菜单
```

```cpp
};
void StuMenu::Show()
{
    cout<<"\n功能菜单:"<<endl;
    cout<<"@@@@@@@@@@@@@@@@@@@@@@@@@@@@@\n";
    cout<<"\t1.选课(1)"<<endl;
    cout<<"\t2.取消选课(2)"<<endl;
    cout<<"\t3.显示本人信息(3)"<<endl;
    cout<<"\t4.根据课程号查询课程(3)"<<endl;
    cout<<"\t5.根据课程名查询课程(5)"<<endl;
    cout<<"\t6.显示所有课程信息(6)"<<endl;
    cout<<"\t7.退出菜单(7)"<<endl;
    cout<<"@@@@@@@@@@@@@@@@@@@@@@@@@@@@@\n";
    cout<<"请输入选择的菜单:";
}

class AdsMenu:public Menu                    //管理员里学生操作菜单
{
public:
    AdsMenu(){};
    void Show();
};
void AdsMenu::Show()
{
    cout<<"\n功能菜单:"<<endl;
    cout<<"~~~~~~~~~~~~~~~~~~~~~~~~~~~~~\n";
    cout<<"\t1.增加学生(1)"<<endl;
    cout<<"\t2.删除学生(2)"<<endl;
    cout<<"\t3.根据学号查询学生信息(3)"<<endl;
    cout<<"\t4.根据姓名查询学生信息(4)"<<endl;
    cout<<"\t5.显示所有学生信息(5)"<<endl;
    cout<<"\t6.退出菜单(6)"<<endl;
    cout<<"~~~~~~~~~~~~~~~~~~~~~~~~~~~~~\n";
    cout<<"请输入选择的菜单:";
}

class AdcMenu:public Menu                    //管理员里课程操作菜单
{
public:
```

```cpp
    AdcMenu(){};
    void Show();
};
void AdcMenu::Show()
{
    cout<<"\n 功能菜单:"<<endl;
    cout<<"~~~~~~~~~~~~~~~~~~~~~~~~~~~~~~~~~~\n";
    cout<<"\t1. 增加课程(1)"<<endl;
    cout<<"\t2. 删除课程(2)"<<endl;
    cout<<"\t3. 根据课程编号查询课程信息(3)"<<endl;
    cout<<"\t4. 根据课程名称查询课程信息(4)"<<endl;
    cout<<"\t5. 显示所有课程信息(5)"<<endl;
    cout<<"\t6. 退出菜单(6)"<<endl;
    cout<<"~~~~~~~~~~~~~~~~~~~~~~~~~~~~~~~~~~\n";
    cout<<"请输入选择的菜单:";
}


void CourseMain()                                  //菜单选择操作
{
    Course *pc=LoadCourse();                       //加载课程文件信息
    Student *ps=LoadStudent();                     //加载学生文件信息
    ElectiveCourse mg(pc,ps);                      //建立学生选课类对象
    char ch;
    SystemMenu smenu;
    AdMenu amenu;
    StuMenu stmenu;
    AdcMenu adcmenu;
    AdsMenu adsmenu;
    string password;
    //定义各菜单对象
    while(1)
    {
        smenu.Show();                              //显示菜单
        ch=smenu.Get();                            //选择菜单
        if(ch=='A'||ch=='a')                       //选择管理员登录
        {
            cout<<"请输入管理员的密码:";
```

```
            cin>>password;
            if(password!="123456")                    //管理员密码不正确
            {
                cout<<"密码不正确,请重新选择功能!!!";
            }
            else                                       //管理员密码正确
            {
                while(1)
                {
                    amenu.Show();                       //显示管理员菜单
                    char ch2=amenu.Get();               //选择管理员菜单
                    if(ch2=='5') break;                 //退出管理员管理功能
                    switch(ch2)
                    {
                    case '1':mg.MaSeCourse();break;    //管理员选课
                    case '2':mg.MaCaCourse();break;    //管理员取消选课
                    case '3': while(1)
                            {
                                adsmenu.Show();         //显示学生操作菜单
                                char c1=adsmenu.Get();  //学生操作菜单选择
                                if(c1=='6') break;      //退出学生操作菜单
                                switch(c1)
                                {
                                case '1': mg.AddStudent();break; //增加学生
                                case '2': mg.SubStudent();break;
                                    //删除学生
                                case '3': int id;
                                    cout<<"请输入需要查找的学生的学号:";
                                      cin>>id;
                                    cout<<endl;
                                    mg.FindStudentId(id);
                                    break;
                                    //根据学号查找学生
                                case '4': mg.FindStudentName();break;
                                    //根据姓名查找学生
                                case '5': mg.ShowStudent();break;
                                    //显示所有学生信息
                                default: cout<<"\n输入错误,请重新输入:";
```

```cpp
                    }
                }
                break;

        case '4':    while(1)
                {
                    adcmenu.Show();                 //显示课程操作菜单
                    char c2=adcmenu.Get();          //课程操作菜单选择
                    if(c2=='6') break;              //退出课程操作菜单
                    switch(c2)
                    {
                    case '1':   mg.AddCourse();break;//增加课程
                    case '2':   mg.SubCourse();break; //删除课程
                    case '3':   mg.FindCourseId();break;
                        //根据课程号查找课程
                    case '4':   mg.FindCourseName();break;
                        //根据课程名查找课程
                    case '5':   mg.ShowCourse();break;
                        //显示所有课程信息
                    default:    cout<<"\n输入错误,请重新输入:";
                    }
                }
                break;
        default:    cout<<"\n输入错误,请重新输入:";
        }
    }
}else
if(ch=='S'||ch=='s')                                //学生用户登录
{
    //用学生的姓名作为密码
    int num;
    string pa;
    cout<<"请输入学生学号和密码:";
    cin>>num>>pa;
    if(mg.FindStudent(num,pa))                      //学生登录成功
    {
        while(1)
```

```cpp
                {
                    stmenu.Show();                          //显示学生菜单
                    char ch3=stmenu.Get();                  //学生菜单的选择
                    if(ch3=='7')  break ;                   //退出学生菜单
                    switch(ch3)
                    {
                    case '1':   mg.StSeCourse(pa);break;    //学生选课
                    case '2':   mg.StCaCourse(pa);break;    //学生取消选课
                    case '3':   mg.FindStudentId(num);break; //根据学号查找学生
                    case '4':   mg.FindCourseId();break;    //根据课程号查找课程
                    case '5':   mg.FindCourseName();break;  //根据课程名查找课程
                    case '6':   mg.ShowCourse();break;      //显示所有课程信息
                    default:    cout<<"输入错误,请重新输入!!";
                    }
                }
            }
            else
                cout<<"学号或密码不正确 ,请重新选择操作!!!"<<endl;
        }else
        if(ch=='Q'||ch=='q')                                //退出系统
        {
            cout<<"谢谢使用学生选课系统,欢迎再次使用!!! \n";
            break;
        }
        else
            cout<<"输入错误,请重新输入!!!";
    }
    SaveStudent(mg.GetStudent());                           //存储学生信息
    SaveCourse(mg.GetCourse());                             //存储课程信息
}
int main()
{
    system("cls");                                          //清屏
    cout<<"\n\t\t 欢迎进入学生选课系统!";
    CourseMain();                                           //调用CourseMain函数,完成选课
    return 0;
}
```